ULTRACENTRIFUGATION OF MACROMOLECULES

Modern Topics

ULTRACENTRIFUGATION OF MACROMOLECULES

Modern Topics

J. W. WILLIAMS

Department of Chemistry
The University of Wisconsin
Madison, Wisconsin

with a Foreword by
STIG CLAESSON

ACADEMIC PRESS   *1972*   *New York and London*

ACADEMIC PRESS, INC.
111 Fifth Avenue, New York, New York 10003

United Kingdom Edition published by
ACADEMIC PRESS, INC. (LONDON) LTD.
24/28 Oval Road, London NW1

LIBRARY OF CONGRESS CATALOG CARD NUMBER: 75-187247

PRINTED IN THE UNITED STATES OF AMERICA

In Memory
Professor The Svedberg
with reverence and affection

Contents

FOREWORD   ix

PREFACE   xi

ACKNOWLEDGMENTS   xiii

LIST OF SYMBOLS   xv

Part 1   POLYDISPERSE SOLUTE SYSTEMS

Chapter I.   **Sedimentation Equilibrium in Polydisperse Nonideal Solutions**

History	5
Theory	6
Sedimentation Equilibrium Experiments	12
Summary	18
References	19

Chapter II.   **Size Distribution Analysis by Ultracentrifugal Methods**

Theory	23
Experiment I. Polystyrene–Cyclohexane at 34.2°C	27
Experiment II. Dextran–Water at 25°C	30
Summary	34
References	35

Part 2   SELF-ASSOCIATION REACTIONS IN PROTEIN SYSTEMS

Chapter III.   Simultaneous Sedimentation and Chemical
Equilibrium in β-Lactoglobulin B,
Chymotrypsinogen A, and Lysozyme Solutions

β-Lactoglobulin B	44
Lysozyme	51
Chymotrypsinogen A	55
Summary	58
References	61

Chapter IV.   Sedimentation Analysis of a Multiple
Myeloma γG-Globulin

Theory	65
Experimental Results	66
Correlation of the Data	70
Summary and Concluding Remarks	74
References	75

Part 3   APPENDIXES

Appendix A.   A Brief Introduction to the Theory of
Ultracentrifugal Analysis

Basic Principles	80
Sedimentation Equilibrium	81
Sedimentation Transport	89
References	96
Bibliography	97

Appendix B.   Molecular Homogeneity and Its Demonstration

Sedimentation Velocity	99
Sedimentation Equilibrium	107
References	112

AUTHOR INDEX	113
SUBJECT INDEX	116

Foreword

For almost half a century Professor J. W. Williams has devoted himself to the study of biopolymers and other macromolecules. He has provided outstanding scientific leadership in this field and has also been a highly successful teacher, inspiring others to do much important and fruitful work. His own work on dielectric properties of molecules and on ultracentrifugation is particularly well known. From a personal friendship of 25 years duration I know that his work on ultracentrifugation is very close to his heart, and therefore I am both pleased and honored to have been asked to write a Foreword to this book, the contents of which were first presented as a series of lectures in the fall term of 1968 during Jack Williams' stay in Uppsala as Nobel Guest Professor.

Professor Williams' interest in ultracentrifugation dates back to 1923 when he was a graduate student and assistant in the chemistry department of the University of Wisconsin. That year Professor The Svedberg had been invited by Professor J. H. Mathews to be visiting professor at Madison. During the voyage across the Atlantic, The Svedberg thought about applying the ultracentrifuge and other methods to the study of colloid systems. In Madison, he was quickly surrounded by an enthusiastic group of scientists—of which young Jack Williams was a member—who began work on ultracentrifugation, diffusion, electrophoresis, preparation of colloids, and related problems. The first primitive ultracentrifuge (the optical centrifuge) was actually built by Svedberg and Nichols in Madison.

Professor Williams spent a happy time in Uppsala, 1934–1935, as International Education Board Fellow. This was the glorious period (beginning 1926 with hemoglobin) during which The Svedberg and his co-workers did their fundamental work on the properties of protein molecules.

After returning to the United States, Professor Williams received a grant from the Rockefeller Foundation for the purchase and development of the ultracentrifuge in the United States, and he was able to install both an oil turbine velocity ultracentrifuge and an electrically driven equilibrium centrifuge built in Uppsala. At first these instruments were used for a series of successful investigations on proteins and related macromolecules. Later, Professor Williams turned his interest to flexible chain polymers. The study of such molecules with the ultracentrifuge had up till then met with considerable difficulty due to several different factors which hindered interpretation of results, the most important of which were the large deviations from ideal behavior and broad molecular weight distributions leading to boundary spreading difficult to deconvolute from diffusion. In a series of fundamental and penetrating papers these effects were studied in detail by Williams and a number of highly competent co-workers, including Baldwin, Fujita, Gosting, Van Holde, and Wales. Of particular importance was their application of the theory of irreversible thermodynamics to these problems.

This work has formed the solid foundation upon which further investigations both by the Madison group and by others have been based. It can truly be said that without Professor Williams' contributions ultracentrifugation would not have become the useful tool of general applicability for the study of chain macromolecules which it is today. This volume presents a review and discussion of the very latest work in this area and, in particular, a summing up of the theoretical description of the problem of sedimentation equilibrium of polydisperse nonideal solutions.

With Professor Williams' famous 1958 paper in *Chemical Reviews* [Williams, J. W., Van Holde, K. E., Baldwin, R. L., and Fujita, H. (1958). *Chem. Rev.* **58**, 715], the proceedings of the 1962 Conference sponsored by the National Academy of Sciences [Williams, J. W. (Ed.) (1963). "Ultracentrifugal Analysis." Academic Press, New York], and of the 1968 Conference sponsored by the New York Academy of Sciences [Yphantis, D. A. (Ed.) (1969). Advances in ultracentrifugal analysis. *Ann. N.Y. Acad. Sci.* **164**, 1–305], this book will form a lasting monoument to his contributions to ultracentrifugal analysis, characterized by deep insight, clarity, and precision.

Stig Claesson
Institute of Physical Chemistry, Uppsala

Preface

This book had its origins in an assignment to me of one of the Nobel Guest Professorships at Uppsala University. It was a great pleasure to return to its wonderful and hospitable Physical Chemistry Institute for another extended visit and to present a series of seminars, so much so that I wanted to make a permanent record of the occasion. I am deeply proud and appreciative of this honor which was bestowed upon me by this University, Fysikalisk—Kemiska Institutionen, and by Professor Stig Claesson.

The general subject of ultracentrifugal analysis had its inception nearly a half century ago. In the course of its development a number of treatises and reviews of the subject have appeared in the scientific literature, and outlines of its basic theories and achievements have found their way into a number of textbooks and encyclopedias.

Two survey statements often came to mind in collating the notes that comprise this volume.

"Perhaps the most striking contribution of physical chemistry to the study of biological macromolecules was the development, in the 1920's, of centrifuges that rotated at high speed (Ultracentrifuges) and that could cause the rapid sedimentation of proteins and nucleic acids" (J. D. Watson, "Molecular Biology of the Gene," 1965, 1970).* From this quotation the inference is

*James D. Watson, "Molecular Biology of the Gene." Copyright © 1965, Benjamin, Menlo Park, California. Reprinted with permission.

clear that of all the physical tools with which biologically oriented scientists were provided in their attempts to unravel complex problems of living matter the ultracentrifuge has been one of great import. It is certainly true that from the use of this equipment by molecular biologists there have come discoveries of great significance and excitement.

"Since no satisfactory theory exists for the excluded volume in a system containing chain molecules of different lengths, any estimate of the second term on the right of equation (4.43) must be considered highly uncertain and a reliable interpretation of the data is possible only when the ploymer is dissolved in a theta-solvent" (H. Morawetz, "Macromolecules in Solution," 1965).* This equation (4.43) is the equivalent of our equation A(25). When the nature of these polymers as macromolecules was realized, it became necessary to have the means to acquire molecular weight or average molecular weight data for them. The ultracentrifuge possesses the unique additional attribute that with its use, in principle at least, the solute heterogeneity can be quantitatively assessed. Whether or not one accepts the prevailing attitude of some polymer physical chemists toward the particular aspect of ultracentrifugal analysis to which reference is made, it will be found that the instrument has had its real triumphs in the domain of polymer investigation, and it promises still more.

The text of the seminars has been substantially revised and brought up-to-date since the time of presentation in the autumn of 1968, however, it remains essentially the same. The subjects, two in organic high polymer chemistry and two in physical biochemistry, were taken from our own more recent research experiences so that what they offer cannot be said to be without bias. The reports are descriptive of experimental researches.

Two appendixes are included along with the account of the researches. In each case an attempt has been made to give an elementary and "in the nutshell" account of the theory necessary for an understanding of the ideas being developed. The equations as written may be lacking in mathematical rigor and elegance, but should be useful in achieving the desired end.

J. W. Williams

*H. Morawetz, "Macromolecules in Solution." Copyright © 1965, Wiley (Interscience) New York. Reprinted with permission.

Acknowledgments

The progress described never could have been made without the earlier collaboration and loyal companionship of a number of talented associates in research who, with me, have considered topics in the general field, some related to biology and medicine and others to technology. Of the particular items now being considered, some of them were developed by recent doctoral candidates at the University of Wisconsin, and a substantial amount of text material has been adapted from their dissertations. Others derive from the participation of project assistants and associates. To all represented, directly or indirectly, I express my heartfelt thanks.

I would like especially to thank Dr. Richard C. Deonier, who has made many useful suggestions for the improvement of the manuscript, Mrs. Barbara J. Eaton, who has patiently and expertly transformed poor handwritten copy into the final typewritten draft; and Mrs. Diane K. Hancock and Mrs. Anne M. Linklater, who have kindly consented to the inclusion of some of their unpublished data.

For my own extended opportunities in research, I wish to thank the Department of Chemistry and the Graduate School of the University of Wisconsin, the Rockefeller Foundation, and certain United States Government Agencies, principally, the National Science Foundation, the National Institutes of Health, and the Office of Naval Research.

Except in two instances the diagrams used to illustrate the text material have been taken from our own publications; most of them have been redrawn or otherwise modified. They are presented by permission of the copyright holders, with year of copyright corresponding to the year of publication. We are indebted to the editors and publishers concerned for this privilege.

Figs. I-1 and I-2	*J. Phys. Chem.* **71,** 2780 (1967)
Figs. I-3 and I-4	*J. Phys. Chem.* **73,** 1448 (1969)
Figs. II-1 and II-2	*J. Phys. Chem.* **68,** 161 (1964)
Figs. II-3–II-6	*J. Phys. Chem.* **58,** 854 (1954)
Figs. III-1 and III-2	*Biochemistry* **11,** 2634 (1972)
Fig. III-3	*Biochemistry* **9,** 4260 (1970)
Figs. III-4 and III-5	*Biochemistry* **8,** 2598 (1969)
Figs. IV-1–IV-8	*J. Biol. Chem.* **241,** 2781, 2787 (1966)
Figs. B-2–B-4	*J. Phys. Chem.* **70,** 309 (1966)

Even my most recent programs have been inspired by what was learned in my very early associations with a great teacher and scientist, Professor The Svedberg, to whose memory this volume is dedicated. The studies depicted originated from his disclosures of the 1920's and 1930's, aided and abetted by such talented individuals as Drs. Björnståhl, Fåhraeus, Lamm, Pedersen, Rinde, and Tiselius. Indeed, present-day research has become what it is because many individuals have persevered in finding solutions to problems, the existence of which had been made a matter of early Uppsala record.

List of Symbols

a	position of meniscus in cell (or r_a)
b	position of cell bottom (or r_b)
c	concentration on volume-based scales, usually g/100 ml but also g/ml
$[c]$	concentration in moles per liter
c_0	original concentration in the cell
c_i	concentration of the ith component
c_a	concentration at the meniscus
c_b	concentration at the cell bottom
$\bar{c}$	numerical average concentration, sedimentation equilibrium $[(c_a + c_b)/2]$
c_{av}	$c_0(r_a/r_H)^2$, sedimentation transport
$c_0{}^*$	Van Holde-Williams concentration parameter
f	molar friction coefficient
f_i	weight fraction of ith species in an equilibrium mixture
f_i^a	apparent value of same in nonideal system
$f(M)$	normalized differential distribution function of molecular weight
g_i	amount of constituent i in the cell
$g(s)$	normalized differential distribution function of sedimentation coefficient
h	cell depth, parallel to axis of revolution
k	proportionality constant
k_s	constant which describes dependence of s on c
n	refractive index
r	radial distance from center of rotation
r_a	radial distance to meniscus (or a)
r_b	radial distance to cell bottom (or b)

xv

$(r_{2m})^2$	position of second moment of boundary gradient curve, sedimentation transport
r_H	position of maximum height of boundary gradient curve, sedimentation transport
s	sedimentation coefficient
s_0	limiting sedimentation coefficient; for solute in two-component system subscript 2 is often omitted
$(s_1)_0$	limiting s value for species 1 in monomer–dimer self-association equilibrium
$(s_2)_0$	corresponding quantity for species 2
t	time
$\bar{v}$	partial specific volume of solute, cm^3/g
x	dimensionless radial distance parameter ($1 - \xi$, where ξ is the corresponding Fujita function)
y_i	activity coefficient of ith species
z	valence of an ion or macroion
A	cell area; factor $(1 - v_2\rho_1)\omega^2/2RT$
B	second virial coefficient (or B_1)
B_{LS}	light-scattering second virial coefficient
D	diffusion coefficient for two-component system, cm^2/sec
D_{ii}	main diffusion coefficients in multicomponent system
D_{ij}	cross-term diffusion coefficients in multicomponent system
F_{ab}	antibody active gamma globulin fragment
F_c	third and inactive gamma globulin fragment
$F(M)$	integral distribution function of molecular weight
$G(s)$	integral distribution function of sedimentation coefficient
G	Gibbs free energy per mole; $\Delta G°$, standard Gibbs free energy change
H	light-scattering factor; enthalpy; $\Delta H°$, standard molal enthalpy change
I	ionic strength
J_i	flow or flux of the ith component
K_2	equilibrium constant for the reaction $2M_1 \rightleftarrows M_2$, dl/g; K_3 for trimerization stage; etc.
K_d	reciprocal of K_2; K_D on molar scale
L	phenomenological coefficient
$(L_{ik})_a$	phenomenological coefficient in multicomponent system for a frame of reference
M	molecular weight
M_1	molecular weight of monomer species
M_2	molecular weight of dimer species; molecular weight of component 2 in ordinary two-component system
M_n	true number average molecular weight over the cell
M_n^a	apparent value of same in nonideal system
M_w	true weight average molecular weight over the cell
M_w^a	apparent value of same in nonideal system
M_z	true z-average molecular weight over the cell
$M_{n(c)}$	number average M which corresponds to position where total solute concentration is c, self-association system
$M_{w(c)}$	weight average M which corresponds to concentration c, self-association system

$M^a_{n(c)}$ and $M^a_{w(c)}$	apparent values of same in nonideal system
M_{wI} and M_{wII}	Goldberg molecular weight parameters
M_0	molecular weight at maximum value of dW/dM versus M curve
M_p	most probable molecular weight in a distrubution
P	pressure
Q	Van Holde-Williams sedimentation equilibrium correction parameter
R	molar gas constant
R_a	$M^a_{w(c)}/M_1$
S	sedimentation coefficient in Svedberg units
T	absolute temperature
U	internal energy per mole
V	cell volume
$\bar{V}$	partial molal volume
W	total weight of solute, normalized
X_k	force on component k
β	nonuniformity coefficient
γ	type of serum globulin molecule
ε	charge per mole of electrons
η	coefficient of viscosity
λ	$(1 - \bar{v}_2\rho_1)\omega^2(b^2 - a^2)/2RT$
μ_i	chemical potential per mole for ith constituent
$\bar{\mu}_i$	total potential per mole for ith constituent
μ_i'	chemical potential per gram
$\bar{\mu}_i'$	total potential per gram
ν	stoichiometric mole number
π	osmotic pressure; ratio of circumference to diameter of a circle, 3.14159. . .
ρ	density of solution, g/cm
ρ_1	density of the solvent
ρ_0	density of dialyzed solvent
τ	excess turbidity
χ	measure of goodness of fit of data
ψ	electrical potential
ω	angular velocity, radians per second
Δ	$(M_w^2\lambda^2/12)(M_z^2/M_w^2 - M_z/M_w)$
Θ	sector angle, ultracentrifuge cell
Π	product
Σ	sum

PART **1**

POLYDISPERSE SOLUTE SYSTEMS

CHAPTER 1

Sedimentation Equilibrium in Polydisperse Nonideal Solutions

As scientists came to realize the significant role of macromolecules in life processes and in technology, new tools for the study of systems containing them had to be envisioned and developed. An important one of these new instruments, the ultracentrifuge, was developed by Svedberg (1). In it convection currents due to temperature inequalities along the column and vibrations due to rotor imbalance are eliminated so that proper mathematical analyses can be made either of the rate of movement or of the equilibrium redistribution of the components of a solution as the rotor revolves at a constant angular velocity, that is, either sedimentation transport or sedimentation equilibrium. Relatively low speeds of a few thousand revolutions per minute are sufficient to give sedimentation rates of the same order as the diffusion rates of proteins and organic high polymers and so to produce a measurable redistribution of concentration at sedimentation equilibrium, while high speeds, 50,000 to 60,000 rpm, may be required to overcome diffusion and produce a moving boundary, the rate of movement of which can be observed in sedimentation velocity experiments.

The first published papers on the ultracentrifuge dealt with analyses of particle size distributions in suspensions of inorganic colloids. The emphasis now has shifted to the organic macromolecules because it was found that a large number of them exists, each of very regular structure.

When the work with the "optical centrifuge" began in 1923 (2), it was the general opinion that giant molecules did not exist. It was believed that the molecular kinetic units of proteins and of organic high polymers in solution were clusters of much smaller molecules, forming particles of undefined mass. Cellulose was thought of as a cyclic tetrasaccharide, with physical association of the smaller units accounting for what we now know to be its high molecular

3

weight. The proteins were considered to be polypeptides of some twenty amino acid residues, aggregated to form reversibly dissociable component systems.

Beginning with the teachings of Svedberg (ultracentrifugal analysis) and of Staudinger (viscometry) such substances were revealed to be macromolecules, large because they contain a huge number of atoms joined together by primary chemical bonds. Following a long and fruitful series of investigations, Svedberg (3) wrote in 1939, "The proteins are built up of particles possessing the hallmark of individuality and therefore are in reality giant molecules. We have reason to believe that the particles in the protein solutions and the protein crystals are built up according to a plan which makes every atom in them indispensable for the completion of the structure."

And it was clearly demonstrated by Staudinger (4) that such substances as polystyrene and natural rubber exist in solution without change in molecular weight, no matter what the solvent. The argument was based upon the premise that viscosity measures molecular length, and in spite of the fact that the original Staudinger molecular weight rule was later found to require modification, the general conclusion has stood the test of time. Thus, such substances as the proteins, polysaccharides, nucleic acids, and the organic high polymers are now described as macromolecules.

In this chapter the sedimentation equilibrium experiment rather than the one based upon sedimentation transport will be discussed. The subject matter will also be restricted by considering linear, soluble synthetic organic high polymers instead of the macromolecules which exist in nature.

In general, the answers to several questions about these macromolecules are sought:

(1) Under what conditions do such macromolecules dissolve as single molecules?

(2) Can their molecular weights be determined?

(3) Are any of them homogeneous with regard to molecular weight and what is the sensitivity of the analytical method by which such homogeneity would be established?

(4) If heterogeneous, is it possible to provide the several average molecular weights and a molecular weight distribution curve for them?

(5) Can the extent of solute–solvent interactions in the solutions be measured?

At least in principle some of these questions have been answered, while progress is being made with others. The sedimentation equilibrium experiment in polymer chemistry is most commonly used to learn about the mass and related properties of a solute and to measure the excess chemical potentials, expressed either as virial coefficients or as activity coefficients.

History

The general mathematical description of the conditions for equilibrium between phases, either in the absence or the presence of an externally applied force (in this instance, that of the ultracentrifuge) was set down by Gibbs nearly a century ago. The kinetic theory approach to the problem came later; it was introduced by Einstein and by von Smoluchowski during 1906–1908. Probably the first experimental observations of a sedimentation equilibrium in an artificial system were made by Perrin at Paris in 1908. These were remarkable experiments, performed with suspensions of gamboge of near uniform size (radius $\cong 0.37$ mμ) in the earth's gravitational field and using a cell which was only 0.1 mm in height. The concentration at the bottom of the cell was roughly twice that at the top at equilibrium. For the earth's atmosphere, another example of a sedimentation equilibrium (but this time in a gaseous polydisperse system), a 50% decrease in pressure requires a height of approximately 5 km for its establishment.

For organic high polymers (and other macromolecules) in solution, it is now obvious that to produce such an equilibrium under reasonable experimental conditions it is necessary to apply an external force field which is much stronger than that due to gravity alone. This was the thought in the mind of Svedberg which led to the development of the ultracentrifuge. In the first machine, the instrument designed to study the heterogeneity of inorganic colloids, the field strength was about 500g. The machine which later came to be known as the Svedberg equilibrium ultracentrifuge is a direct successor to this first "optical centrifuge," so-called, the main differences being that the rotating part or rotor which contains the solution cell is now contained in a gas chamber which, in turn, is immersed in a thermostat.

The first real tests of the sedimentation equilibrium experiment in application to organic high polymers were made almost simultaneously by Lansing and Kraemer (5) in 1935 and by Signer and Gross (6) in 1934. These beginnings, promising as they were, were not immediately developed and enriched, and it was not until well after World War II that polymer chemists began a more intensive use of the ultracentrifuge. For the most part it was the sedimentation transport experiment to which they turned, but because of the flexible character of the polymer molecules, the compressibility of the solvent, and so on they had to accept a much diminished return for their efforts as compared to that which accrued to the protein physical chemist who had for study rigid, globular molecules in aqueous solutions. But in spite of the relatively much greater success of the physical biochemists, one still has to feel that more would have been achieved for all, had the potentialities of the sedimentation equilibrium experiment been given earlier and more serious consideration. It will be our purpose to try to justify this statement.

Theory

The basic idea in the derivation of the essential working equations for sedimentation equilibrium is that in the presence of the externally applied ultracentrifugal field and at equilibrium, the total potential of any constituent is constant in all phases, i.e., at each radial distance in the cell. The equations which are required for the interpretation of the experimental data may be quite simple or they may be algebraically complex, depending upon the nature of the system. The traditional expressions for the equilibrium condition apply only to the case of an ideal, isothermal, incompressible solution made up of two monodisperse components, both of them electrically neutral molecules. Complications arise when the partial specific volumes and densities depend on pressure, when there are solvent–solute interactions, when the solute is polydisperse, when it exists in the form of macroion with counterions, and so on. Numerous equations which take such additional effects into account are to be found in the literature. Perhaps the most useful source of information is still the Fujita monograph, but there are several other accounts, largely review articles, where the several more involved equations are also set down. In Appendix A an attempt has been made to suggest the routes by which the basic essential working equations are obtained, not only for sedimentation equilibrium, but also for sedimentation transport.

SEDIMENTATION EQUILIBRIUM OF POLYDISPERSE CONCENTRATION-DEPENDENT SYSTEMS

The sedimentation equilibrium experiment has been described by Van Holde as "the most versatile of all methods, and in most cases, the most accurate as well." (In this statement the comparison was with the thermodynamic properties which are found from osmotic pressure and light-scattering experiments.) Yet, as has just been indicated, the sedimentation equilibrium approach has not found the same general acceptance in organic high polymer chemistry that it has received by the protein physical chemist. There are several reasons for this situation, the most important one being that the theoretical equations for the description of the equilibrium in polydisperse, nonideal systems are relatively complex. Indeed, the belief has become widespread that interpretable measurements can be made *only* at the Flory temperature—that is, in pseudo-ideal solutions, when the solutes are synthetic organic high polymers. We have never been quite willing to accept this proposition, and a record of efforts to modify it is given here.

Wales *et al.* in 1946–1951 (7) were interested in the sedimentation equilibrium of polydisperse nonideal solutes but there was some difficulty in the interpretation of the ultracentrifuge records. The data showed that the apparent weight average molecular weight, M_w^a, computed as if the system were ideal in thermodynamic behavior, depended *both* upon polymer concentration and rotor speed. Insofar as concentration-dependent effects are concerned, it was reasoned that thermodynamically, the measurement of sedimentation equilibrium is equivalent to the determination of osmotic pressure, a good place from which to start. (Actually in 1944, G. V. Schulz in Mainz had made use of a substantially equivalent analysis, but because of wartime conditions, his work was unknown to us until after 1946.)

One important milestone was passed in 1951 (*8*) when the sedimentation equilibrium experiment became a self-contained one; it was no longer necessary to resort to separate osmotic pressure experiments for the determination of a second virial coefficient. And, Wales *et al.* (*8*), as suggested in Appendix A, could conclude that "for a small cell, or its equivalent, a not too great separation of components," one may write

$$\frac{1}{M_w^a} = \frac{1}{M_w} + Bc_0$$

where M_w^a is the apparent weight average molecular weight, M_w is the true weight average molecular weight, and c_0 is the initial solute concentration in grams per 100 ml. It was an empirical equation, but one which was written with remarkable intuition.

However, in the case of the polydisperse nonideal system, the situation is quite complex, because at any fixed point in the cell the distribution of molecular weights at equilibrium will differ from that in the original solution; furthermore, the correction for the solute–solvent interactions will be dependent not only upon the particular molecular weight distribution (everywhere different along the column) but also upon the total concentration at the given point. Thus, the early treatments of Schulz and of Wales, steps along the way to be sure, are now recognized to include approximations which are known to be only roughly correct; also they did not provide the means to eliminate the effect of rotor speed variations. It was not until 1953 that Goldberg (*9*) furnished the first completely general mathematical analysis for a multicomponent system at equilibrium when solution nonidealities are present. Further progress could then be made.

Considering momentarily the two-component system (neutral molecules) at equilibrium, we have

$$\frac{1}{c_2}\frac{dc_2}{dr} = \frac{M_2(1 - \bar{v}_2\rho)\omega^2 r}{RT} \qquad \text{(ideal)} \qquad \text{(I-1a)}$$

$$= \frac{M_2^a(1 - \bar{v}_2\rho)\omega^2 r}{RT} \qquad \text{(nonideal)} \qquad \text{(I-1b)}$$

Then,

$$\int_a^b \frac{dc_2}{dr}\,dr = \frac{M_2(1 - \bar{v}_2\rho)\omega^2}{RT}\int_a^b rc(r)\,dr \tag{I-2}$$

With the use of Equation (A-3)* and the assumption that in a dilute system the solution density, ρ, may be approximated by that of the solvent, ρ_1,

$$\frac{\Delta c}{c_0} = \frac{c_b - c_a}{c_0} = \frac{M_2(1 - \bar{v}_2\rho_1)\omega^2(b^2 - a^2)}{2RT} = \lambda M_2 \quad \text{(ideal)} \tag{I-3a}$$

$$= \lambda M_2^{\text{a}} \quad \text{(nonideal)} \tag{I-3b}$$

So, in the nonideal solution, $\lambda = \dfrac{(1 - \bar{v}_2\rho)\omega^2(b^2 - a^2)}{2RT}$

$$M_2^{\text{a}} = (c_b - c_c)/\lambda c_0 \tag{I-4}$$

where M_2^{a} is the apparent molecular weight of the solute component and c_b and c_a are the solute equilibrium concentrations at the bottom and at the meniscus of the cell, respectively. Then, with Equation (A-15),

$$\frac{1}{M_2^{\text{a}}} = \frac{1}{M_2} + \tfrac{1}{2}B_2'(c_b + c_a) \cong \frac{1}{M_2} + B_2\bar{c}$$

where $\bar{c}$ is the average concentration over the cell at equilibrium.

According to Equation (A-16), extrapolation of $1/M_2^{\text{a}}$ to infinite dilution in the variable $\bar{c}$ yields $1/M_2$ as the ordinate intercept and the second virial coefficient B_2 as the limiting slope. Only neutral molecules are involved, so the second virial coefficient and the activity coefficient y for the two-component systems can be related by the equation, $\ln y = B_2M_2c + \ldots.$ Equations (I-3) are the basic statements for the "low-speed" sedimentation equilibrium method, the one which has become standard in practice in the author's laboratory. The descriptive term is not a good one, because constant solution depth is also involved. It has come into use by way of contrast to the "high-speed" (or "meniscus depletion") mode of operation, first suggested by Wales, Adler, and Van Holde and more recently extensively exploited by Yphantis.

The analysis can be extended to a polymer system in which the solute consists of components differing in molecular weight. The equations which describe the sedimentation equilibrium, with the $\bar{\mu}_i$ written on a per *gram* basis, are (9):

$$(1 - \bar{v}_i\rho)\omega^2 r = \sum_{k=2}^{q}\left(\frac{\partial\mu_i}{\partial c_k}\right)\frac{dc_k}{dr} \quad i = 2, \ldots, q \tag{I-5}$$

* Equations with an "A" are to be found in Appendix A.

where the several quantities have their conventional significance, with q representing the total number of components. The task at hand is to put these equations into a form so that a series of experiments performed at several low rotor speeds and at several low concentrations can provide the information which is necessary to compute both the weight average molecular weight and the light-scattering second virial coefficient (10).

For a solution which is sufficiently dilute in all solutes, we write

$$\ln y_i = M_i \sum_{k=2}^{q} B_{ik} c_k + \cdots \tag{I-6}$$

Then, with the use of the dimensionless parameter (11, 11a)

$$x = \frac{r^2 - r_a^2}{r_b^2 - r_a^2}$$

and the definition of λ as provided by Equation (I-3), it can be shown that

$$\lambda M_i c_i = \frac{dc_i}{dx} + c_i M_i \sum_{k=2}^{q} B_{ik} \frac{dc_k}{dx} + \bar{v} \sum_{k=2}^{q} c_k \frac{dc_i}{dx} \tag{I-7}$$

in which all the $\bar{v}_i$ have been assumed to be constant and where $\bar{v}$ is defined as $\Sigma c_i \bar{v}_i / \Sigma c_i$. This expression describes the concentration gradient of a typical polymeric species at sedimentation equilibrium.

With the use of two approximations (which become exact as $\lambda \to 0$)

$$\frac{dc_k}{dx} \cong \lambda M_k c_k \quad ; \quad c_k(x) \cong \frac{(c_k)_0 \lambda M_k \exp \lambda M_k x}{(\exp \lambda M_k) - 1}$$

in Equation (I-7), followed by expansion of the exponential, we obtain

$$\lambda M_i c_i = \frac{dc_i}{dx} + \lambda M_i c_i \sum_{k=2}^{q} (c_k)_0 M_k \left(B_{ik} + \frac{\bar{v}}{M_k} \right) + \text{higher terms in } \lambda^2 \tag{I-8}$$

$$B_{ik} = \frac{1}{M_i} \left(\frac{\partial \ln y_i}{\partial c_k} \right)_{T,P,c_j}$$

On integration, summation over all solute species ($k = 2, \ldots, q$), and rearrangement we arrive at the very important relationship (11a)

$$\frac{\Delta c}{c_0} = \lambda (M_w - B_{LS}^1 c_0 + \text{higher terms in both } \lambda^2 \text{ and } c_0^2) \tag{I-9}$$

where

$$B_{LS}^1 = M_w^2 B_{LS}$$

$$B_{LS} = \frac{1}{M_w^2} \sum_{i=2}^{q} \sum_{k=2}^{q} f_i f_k M_i M_k \left(B_{ik} + \frac{\bar{v}}{M_k} \right)$$

$$= \text{light-scattering second virial coefficient}$$

$$f_i = (c_i)_0/c_0 \qquad f_k = (c_k)_0/c_0$$

Equation (I-9) will be seen to reduce to the common form for an ideal system, namely, $\Delta c/c_0 = \lambda M_w$.

In reciprocal form, Equation (I-9) becomes

$$\frac{\lambda c_0}{\Delta c} = \frac{1}{M_w} + B_{LS}c_0 + \cdots \tag{I-10}$$

It is the common expression for the analysis of the experimental data.

There have been numerous examples in the literature in which investigators have applied an equation similar to Equation (A-16) in the interpretation of data for nonideal solutions of polydisperse solutes, simply by replacing M_2 by M_w and regarding $B_2 = B$ as a parameter which may be comparable to the second virial coefficient obtainable from light-scattering measurements. But such empirical procedures still require theoretical justification, and we shall consider, briefly to be sure, under what approximations such action becomes permissible. A more complete analysis of the situation has recently been presented by Deonier and Williams (12).

In 1953, Van Holde and Williams (13) recognized that an extrapolation of $(M_w^a)^{-1}$ to infinite dilution might be better made by using a redefined concentration variable instead of the initial concentration c_0. In condensed form their starting equation [cf. Equation (I-7)] is

$$\lambda M_{w(r)}c = dc/dx + M_{w(r)}Bc(dc/dx) + \ldots$$

It contains the assumption that $B_{ik} = B$ with $i,k = 2, \ldots, q$. With the definition

$$\left(\frac{1}{\lambda c} \right) \frac{dc}{dx} = M_{w(r)}^a$$

this equation can be written as

$$M_{w(r)} = M_{w(r)}^a [1 + BM_{w(r)}^a c + \cdots] \tag{I-11}$$

Then, since

$$M_w = \frac{\displaystyle\int_0^1 M_{w(r)}c \, dx}{\displaystyle\int_0^1 c \, dx}$$

it was possible to derive the expression

$$(M_w^a)^{-1} = (M_w)^{-1} + Bc_0^*Q \tag{I-12}$$

in which

$$c_0^* = \frac{\left[\int_0^1 (M_{w(r)}^a)^2 c^2 \, dx\right]\left[\int_0^1 c \, dx\right]}{\left[\int_0^1 M_{w(r)}^a c \, dx\right]^2}$$

and Q is a complicated factor which may not differ significantly from unity. Thus, $(M_w^a)^{-1}$ may be plotted against c_0^* to yield $(M_w)^{-1}$ as the ordinate intercept.

In more recent years Fujita has returned to the problem of finding suitable approximations to the general sedimentation equilibrium equation which applies to polydisperse nonideal solutions. In 1959 (14) he developed the statement

$$(M_w^a)^{-1} = (M_w)^{-1} + Bc_0\left(1 + \frac{\lambda^2 M_z^2}{12} + \cdots\right)$$

in which M_z is the "z" average molecular weight. This equation derives from an expanded form of Equation (I-7) to give not only the intercept $(M_w)^{-1}$ but also the limiting slope of the $(M_w^a)^{-1}$ versus c curve. Its derivation is to be found in the Fujita monograph (11). A decade later, he (15) presented another approximation, in the form

$$(M_w^a)^{-1} = (M_w)^{-1} + B\bar{c}[1 + \Delta] \qquad \text{(I-13)}$$

In these approximations $B = B_{ik}$; $\bar{c}$ (as before) is the average concentration over the cell column; and Δ, a quantity for which an exact mathematical statement is written in the original article, is small when λ is sufficiently small or when the molecular weight distribution is sufficiently sharp.

All three treatments, the one by Van Holde and Williams, and the other two by Fujita, started with approximate forms of the same differential equations, and it seemed surprising that they should appear to give different working equations. It was possible to show (12a), that under proper experimental conditions, that is *when the product λM_w is sufficiently small*, the three approximations are substantially equivalent, since

$$c_0[1 + (\lambda^2 M_z^2/12)] = \tfrac{1}{2}(c_b + c_a)[1 + \Delta] = c_0^* Q \qquad \text{(I-14)}$$

Goldberg, (9) using again the same starting equation, had derived the expression

$$M_w = M_{wI} + BM_{wII}$$

with terms of higher order in B being neglected. Mathematical definitions of M_{wI} and M_{wII} are to be found in the original article. Upon rearrangement this statement becomes $(M_{wI})^{-1} = (M_w)^{-1} + B[M_{wII}/M_w M_{wI}]$.

Deonier (*12a*) has demonstrated that the concentration variable, in brackets is equal to $\bar{c}(1 + \Delta)$, so that the Goldberg equation is also equivalent to the approximations of Van Holde and Williams and of Fujita. However, from the text of the article it is clear that Goldberg was not thinking in terms of a variable for extrapolation, and the report is not here considered.

We have then justification for the more general use of the average concentration, $\bar{c}$, as a useful variable for the extrapolation of $(M_w^a)^{-1}$ data to infinite dilution, when $\lambda M_w \leq 1$, to be sure. This variable is somewhat superior to the initial concentration as used by Wales, Adler, and Van Holde (*8*) because it automatically accounts for some of the dependence of M_w^a upon λ. The requirement that $\lambda M_w \leq 1$ is met by conducting the experiment at low rotor speeds and with short solution columns, a more-up-to-date statement of what Wales *et al.* had set down as being the condition for successful use of their empirical equation.

Sedimentation Equilibrium Experiments

Of several experimental efforts which provide tests of these equivalent approximations, three are discussed here. In each of these sufficient background theory was available in the literature at the time the experimental work was performed to design the procedure and to guide the most advantageous plotting of the data in order to obtain true weight average molecular weight and second virial coefficient data.

The first (*16*), the method of Van Holde and Williams, was applied to the study of polyisobutylene in isooctane, a "good" solvent, at 25°C, and the resulting extrapolated values for $(M_w^a)^{-1}$ were compared to the corresponding quantities for this polymer determined in ethyl *n*-heptanoate at 34°C, which is the Flory temperature for this system. Discrepancies as large as 25% were detected in the higher molecular weight range, 278,000–588,000 g/mole. It was concluded that the method of Van Holde and Williams, using the concentration variable c_0^* for the plots, would be inaccurate and unsuitable except when applied to solutes of relatively low molecular weight.

In an additional experimental test of the applicability of the new methods for plotting the sedimentation equilibrium data, Albright and Williams (*17*) utilized solutions of two polystyrenes dissolved in toluene at 20°C and a polyisobutylene dissolved in cyclohexane, also at 20°C; both solvents answer the description "good" at this temperature. These particular systems were selected because both weight average molecular weight and second virial coefficient data were independently available in the literature, for later comparison. It is true, however, that both the higher molecular weight polystyrene and the polyisobutylene samples were fractionated materials, and that the second polystyrene, though polydisperse, was of low molecular weight. Thus, in a very real sense, a really rigorous test of the efficacy of the Albright-Williams method was not provided. The experiments were intended as a step along the way.

To guide the treatment of the experimental data Equations (I-9) and (I-10) were used. The route by which the correction of the actual data for $M_w{}^a$, both for variations in λ (i.e., variation of rotor speed at constant solution column depth) and in concentration is now suggested. With reference to Equation (I-9), as $\lambda \to 0$, we have

$$\lim_{\lambda \to 0} \left(\frac{\Delta c}{\lambda c_0} \right) = M_w - B_{LS} M_w{}^2 c_0 \qquad \text{(I-9a)}$$

Two plots are involved; the first is

$$\frac{\Delta c}{c_0} \text{ versus } \lambda$$

$\Delta C = C_b - C_a$

$C_0 = $ initial solute conc.

Whenever the approximations made in the derivation of Equation (I-9) are permissible, a near-linear plot will result, and the limiting slope will be, in turn, a linear function of c_0.

The second plot is

$$\lim_{\lambda \to 0} \left(\frac{\Delta c}{\lambda c_0} \right) \text{ versus } c_0$$

This plot will have an intercept equal to the true weight average molecular weight, M_w, and a limiting slope equal to the product of $M_w{}^2$ and the light-scattering second virial coefficient B_{LS}.

Actually, the reciprocal plot, guided by Equation (I-10), is the usual one for the treatment of experimental data of this kind; indeed, it is in general to be preferred.

The procedure suggested, then, is to conduct a series of experiments at different low initial concentrations, using for each concentration a series of λ values. The rotor speeds and solution column heights must be so selected that the corresponding values of λ are less than $1/M_w$.

The necessary data were obtained by using a Spinco Model E analytical ultracentrifuge. For the lower solution concentrations Rayleigh interference optics were employed, in this way to permit more accurate evaluation of the concentration differences over the cell column. Also, in many of the experiments the 30 mm cell provided a longer optical path through the cell than the more conventional 12 mm cell. The heights of the solution columns were adjusted to 2.75 mm. Rotor speeds were always selected so that the values of λ were of the magnitude $1/M_w$. The information required is the quantity $\Delta c = c_b - c_a$ over the cell column at equilibrium as a function of λ and as a function of the initial polymer concentration, c_0.

The $\Delta c/c_0$ values from the experiment for the several initial concentrations of the polystyrene designated as S-108 dissolved in toluene at 20°C are plotted as a function of λ in Figure I-1. The slopes in these plots decrease with increasing solute concentration, an indication of concentration dependence behavior.

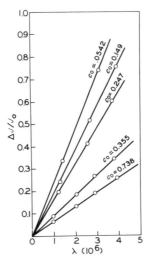

Figure I-1. Plot of the quantity $\Delta J/J_0$ as a function of λ of the system polystyrene S-108 and toluene at sedimentation equilibrium. (Temperature = 20°C.) Redrawn from Albright and Williams (*17*). Reprinted with modification from *J. Phys. Chem.* **71**, 2780 (1967). Copyright (1967) by the American Chemical Society. Reprinted by permission of the copyright owner.

Then, in Figure I-2, Curve A, graphs of the *limiting slopes* of the $\Delta J/J_0$ ($= \Delta c/c_0$) versus λ plots of Figure I-1 versus polymer initial concentration, c_0, are presented. It is the limiting values of these slopes as λ approaches zero which are to be graphed against the initial concentrations, because according to Equation (I-9), it is only in this limit that the experiment will yield the true light-scattering second virial coefficient. In this same figure, Curve B presents the reciprocal plot, guided by Equation (I-10).

Although the M_w data obtained from the two intercepts are in excellent agreement and well within experimental error, we repeat that the reciprocal plot is generally to be preferred. For the evaluation of the light-scattering second virial coefficient, however, the choice may become more involved, but time does not permit the amplification of this statement.

In Table I-1 a summary of the data for the three polymer–solvent–temperature combinations is presented. To complete the comparison of our sedimentation equilibrium thermodynamic quantities with those already recorded

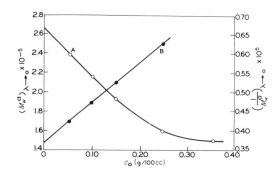

Figure I-2. *Curve A.* Plot of the quantity $\lim\limits_{\lambda \to 0}\left(\dfrac{1}{\lambda}\dfrac{\Delta c}{c_0}\right)$ as a function of initial concentration, c_0, for the polystyrene S-108, toluene system. *Curve B.* Plot of the quantity $\lim\limits_{\lambda \to 0}\left(\dfrac{\lambda c_0}{\Delta c}\right)$ versus c_0 for the same system. This is the conventional reciprocal plot, but one in which the effects of variation of rotor speed have been eliminated. Redrawn from Albright and Williams (*17*). Reprinted with modification from *J. Phys. Chem.* **71**, 2780 (1967). Copyright (1967) by the American Chemical Society. Reprinted by permission of the copyright owner.

TABLE I-1

Data for Molecular Weight and Light-Scattering Second Virial Coefficients Obtained at Sedimentation Equilibrium

System	M_w/M_n	M_w	M_w (present data)	$B_{LS}/2 \times 10^{4\,a}$	$B_{op} \times 10^{4\,b}$
Polystyrene S-108 + toluene	1.08	267,000	267,000 (A) 271,000 (B)	3.75	4.0
Polyisobutylene G-13 + cyclohexane	1.01–1.02	74,000 (M_v)	75,000	7.0	7.4
Oligostyrene + toluene	1.75	5,800^c (D.A.A.)	5,840	9.5	10.2

^{a} Units for B_{LS} are ml mole g^{-2}. Data obtained by using Equation (I-9), which we prefer for this computation.

^{b} Calculated from Flory-Krigbaum equations.

^{c} Owing to the fact that apparent M_w data in an earlier study [S. I. Klenin, H. Fujita, and D. A. Albright, *J. Phys. Chem.* **70**, 946 (1966)] were not corrected to $\lambda \to 0$ and because of a slight computational error, the M_w value for this oligostyrene appears as 5600.

in the literature we have included a column which gives the osmotic pressure second virial coefficients for polymers with number average molecular weights which compare with the corresponding estimated M_n values of our samples. These data were computed by using a Flory-Krigbaum relationship between molecular weight and second virial coefficient. It has been assumed that $B_{LS} = 2 B_{op}$, the light-scattering and osmotic pressure coefficients, respectively, a relationship which is strictly true only if the polymer is sharply fractionated, as was indeed the case for two of the samples.

The third of the experimental researches for consideration is that of Utiyama, Tagata, and Kurata (18). The authors were "concerned with the establishment of a standard experimental method for the sedimentation equilibrium" in application "to both polydisperse and monodisperse nonideal polymer solutions of molecular weights ranging from several tens of thousands to several millions." Their publication (18) is an excellent source of information for the subject matter of its title: "Determination of Molecular Weight and Second Virial Coefficient of Polydisperse Nonideal Polymer Solutions by the Sedimentation Equilibrium Method."

Of main concern is their use of λ and, apparently unknowingly, of the very near equivalent of $\bar{c}$ as experimental variables in the interpretation of their data, but they also discussed the application of the ultracentrifuge to studies of polydisperse nonideal solutes at constant values of λ with initial solute concentration, c_0, as the concentration parameter. Two samples of essentially monodisperse polystyrenes and four blends of these materials were dissolved in 2-butanone at 25°C and examined in the ultracentrifuge at sedimentation equilibrium. Solution column heights of 1.5 mm used. The

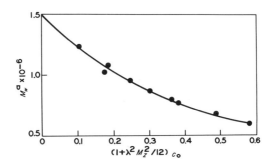

Figure I-3. This curve corresponds to Curve A of Figure I-2. The solute is a blended polystyrene, the solvent is 2-butanone, the temperature is 25°C. The abscissas are now closely equivalent to the average concentration, $\bar{c}$, a parameter which compensates for the effect of rotor speed on $M_w{}^a$. Redrawn from Utiyama et al. (18). Reprinted with modification from *J. Phys. Chem.* **73**, 1448 (1969). Copyright (1969) by the American Chemical Society. Reprinted by permission of the copyright owner.

apparatus was equipped with a Rayleigh interference optical system. In all cases, the weight average molecular weight, M_w, obtained was correct to within $\pm 2\%$ irrespective of the degree of polydispersity of the sample.

A point of interest is their display of plots for both $M_w{}^a$ and $(M_w{}^a)^{-1}$ versus the concentration variable $[1 + (\lambda^2 M_z{}^2/12)]c_0$ for one of the blended polystyrenes. These plots, shown here as Figures I-3 and I-4, are substantially

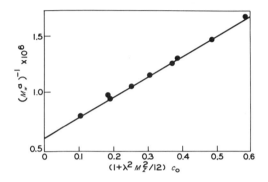

Figure I-4. Plot showing the linear dependence of the quantity $1/M_w{}^a$ versus $[1 + (\lambda^2 M_z{}^2/12)]c_0$ (or $\bar{c}$). The data are taken from Figure I-3. Redrawn from Utiyama *et al.* (*18*). Reprinted with modification from *J. Phys. Chem.* **73**, 1448 (1969). Copyright (1969) by the American Chemical Society. Reprinted by permission of the copyright owner.

the same as the Albright-Williams Curves A and B, respectively, presented in Figure I-2, from which the effects of rotor speed variations have already been eliminated. In Equation (I-14), we have indicated that, except for a small correction term,

$$\left(1 + \frac{\lambda^2 M_z{}^2}{12}\right)c_0 \cong \bar{c} = \frac{c_b + c_a}{2}$$

when $\lambda M_w \sim 1$ (or the molecular weight distribution is sufficiently narrow). Intercept and initial slope of the plot $(M_w{}^a)^{-1}$ versus $\bar{c}$ provide values for M_w and B_{LS}, respectively. As already indicated, the use of the variable $\bar{c}$ automatically accounts for some of the dependence of $M_w{}^a$ upon λ. We believe the original Van Holde-Williams variable $c_0{}^*$ to be superior to either $\bar{c}$ or $[1 + (\lambda^2 M_z{}^2/12)]c_0$ for the use of consideration, but here again, the difference is small and the present experimental accuracy is such that there is hardly a basis for real choice. Perhaps it should be noted that the plots of $(M_w{}^a)^{-1}$ versus $[1 + (\lambda^2 M_z{}^2/12)]c_0$ to obtain M_w and B_{LS} could not ordinarily be used, since data for M_z would not be available.

One question remains. If the concentration variable used by Mandelkern et $al.$ (16), namely c_0*, differs so slightly from $\bar{c}$ and $[1 + (\lambda^2 M_z^2/12)]c_0$, why is it that these investigators failed to obtain reliable M_w data? With reference to these results Utiyama et $al.$ remark, "The difficulty in the extrapolation of $1/M_{app}$ to $c_0 = 0$ reported by Mandelkern et $al.$, for a high molecular-weight sample of polyisobutylene in iso-octane is simply due to the fact that the value of $M\lambda$ used in their experiment was too large." Actually, in the Mandelkern et $al.$ report there is an insufficient description of their experimental conditions, but there is internal evidence that long solution columns were used in the experiments; thus it does seem probable that λM_w was too large, perhaps even by a factor of 10, for the treatment given to the data. In the interpretation of these experiments, the Van Holde-Williams approximation was utilized but, as has been remarked, it is not universally applicable. Short solution columns and low operating rotor speeds to give $\lambda \sim 1/M_w$ are required. On the basis of the success of the Albright-Williams and Utiyama et $al.$ studies, we have to conclude that had Mandelkern et $al.$ used proper experimental conditions, which had already been suggested empirically by Wales et $al.$ in 1951, they probably could and would have obtained a close approximation to the true M_w by using the available theory.

Summary

It has been amply indicated that the interpretation of the data from the sedimentation equilibrium experiment for polydisperse nonideal solutions is considerably more involved than the treatment of light-scattering and osmotic pressure information. The difference lies in the fact that a system in which the distribution of solute molecules is not uniform along the cell is under consideration. However, there are certain advantages inherent in the sedimentation equilibrium experiment and it has been of interest to us and to others to study the operational conditions under which accurate data for the weight average molecular weight of the dissolved polymer can be obtained along with reasonably good second virial coefficient information.

To be sure, the literature of the subject is not a source of encouragement to proceed. There is no satisfactory theory by which the excluded volume in the system which contains chainlike molecules of varying length can be estimated. Thus an independent evaluation of the activity coefficient term in the basic equation descriptive of the equilibrium is seemingly out of the question. It is not at all difficult to find recorded statements to the effect that reliable interpretation of the data is possible only when the experiment is performed at the Flory temperature, where conditions of pseudo-ideality prevail.

Fujita recognized that the early treatments of the problem were approximate to the extent that there was neglected the dependence on molecular weight of the pair-thermodynamic interaction parameters which appear in the first term of the virial expansion for $\ln y_i$ of each solute in the polydisperse system. This effect was included by him in the general theory in an ingenious way to permit the further experimental development of the subject, which began with Fujita, Linklater, and Williams (19). But it was not until the advances made by Adams (20), Osterhoudt (10), and Albright (17) that a quite satisfactory general method for the estimation of M_w and B_{LS} really became available. They showed how a series of experiments for a polymer in a good solvent at several low ultracentrifugal field strengths could be evaluated to give M_w and the light-scattering virial coefficient B_{LS}. Their procedures have been employed, first by Albright and Williams, and then by Utiyama et al. (18). We have presented a condensed account of the record.

Our own experiments are open to the objection that they do not provide a general and rigorous test; it was the intention to move forward slowly and deliberately. The data of Utiyama et al. remove any such limitation for they appear to establish their claim that "the weight average molecular weight of the dissolved polymer can be obtained within experimental error of $\pm 2\%$ irrespective of the degree of polydispersity." Of course, the error involved in the determination of the second virial coefficient is much affected by the type of molecular weight distribution, but even here, it may not be greater than 5% with the usual polydisperse polymers. It is thus considered proper to state that the ultracentrifuge is rapidly taking a place already established for the osmometer and the light-scattering photometer as a means of study of the thermodynamic properties of polymer solutions, and extensions to even more complicated systems may be anticipated.

References

1. T. Svedberg and K. O. Pedersen, "The Ultracentrifuge." Oxford Univ. Press (Clarendon), London and New York, 1940.
2. T. Svedberg and J. B. Nichols, J. Amer. Chem. Soc. 45, 2910 (1923).
3. T. Svedberg, Proc. Roy. Soc., Ser. A 170, 40; Ser. B 127, 1 (1939).
4. H. Staudinger, Chem. Ber. 53, 1073 (1920); 57, 1203 (1924).
5. W. D. Lansing and E. O. Kraemer, J. Amer. Chem. Soc. 57, 1369 (1935).
6. R. Signer and H. Gross, Helv. Chim. Acta 17, 59, 335, and 726 (1934).
7. M. Wales, M. M. Bender, J. W. Williams and R. H. Ewart, J. Chem. Phys. 14, 353 (1946); M. Wales, J. Phys. Colloid Chem. 52, 235 (1948); M. Wales, J. W. Williams, J. O. Thompson and R. H. Ewart, ibid. p. 983; M. Wales, ibid. 55, 282 (1951).
8. M. Wales, F. T. Adler and K. E. Van Holde, J. Phys. Chem. 55, 145 (1951).
9. R. J. Goldberg, J. Phys. Chem. 57, 194 (1953).
10. H. W. Osterhoudt and J. W. Williams, J. Phys. Chem. 69, 1050 (1965).

11. H. Fujita, "The Mathematical Theory of Sedimentation Analysis." Academic Press, New York, 1962.
11a. T. H. Donnelly, *Ann. N.Y. Acad. Sci.* **164**, 147 (1969).
12. R. C. Deonier and J. W. Williams, *Proc. Nat. Acad. Sci. U.S.* **64**, 828 (1969).
12a. R. C. Deonier, Ph.D. Dissertation, University of Wisconsin (1970).
13. K. E. Van Holde and J. W. Williams, *J. Polym. Sci.* **11**, 243 (1953).
14. H. Fujita, *J. Phys. Chem.* **63**, 1326 (1959).
15. H. Fujita, *J. Phys. Chem.* **73**, 1759 (1969).
16. L. Mandelkern, L. C. Williams and S. G. Weissberg, *J. Phys. Chem.* **61**, 271 (1957).
17. D. A. Albright and J. W. Williams, *J. Phys. Chem.* **71**, 2780 (1967).
18. H. Utiyama, N. Tagata and M. Kurata, *J. Phys. Chem.* **73**, 1448 (1969).
19. H. Fujita, A. M. Linklater and J. W. Williams, *J. Amer. Chem. Soc.* **82**, 379 (1960).
20. E. T. Adams, Jr., Dissertation, University of Wisconsin (1962).

CHAPTER II

Size Distribution Analysis by Ultracentrifugal Methods

The evaluation of polydispersity is a significant part of ultracentrifugal analysis. With a nonuniform solute the centrifugal field causes a partial separation, a physical fractionation, of the component parts in the solution cell, and opportunities are provided to determine the distribution of sedimentation coefficients, or in the case of the sedimentation equilibrium experiment, to give an absolute molecular weight distribution. And there are instances in which it may be possible to combine the results of the two experiments to learn about molecular forms and reaction mechanisms.

As early as 1928, H. Rinde (1) described the equations by which it is possible to calculate the distribution curve for colloidal particles in suspension, either from a sedimentation equilibrium curve or from the distributions of the concentrations in the cell after a given time of sedimentation transport. Those of us who have followed have profited greatly by the use of his ideas and the extensions of his equations. Very shortly after the publication of Rinde's dissertation it was noted by Svedberg and his collaborators that in polydisperse macromolecular solutions the apparent diffusion "constant" calculated from the boundary or boundary gradient versus distance curve in sedimentation velocity experiments increased with time; this behavior was properly attributed to the heterogeneity of the solute. In the Uppsala theses of Gralén (1944) and Jullander (1945) are found more quantitative studies of the broadening of such boundary gradient curves with distance in the cell as a measure of polymolecularity. (In principle, the sedimentation velocity approach used by us answers this description, but the procedure is different.)

Thus, the evaluation of the heterogeneity of a macromolecular system with respect to the sedimentation coefficient first really became feasible at the close of World War II. In our attempts to advance the subject we have extended the basic formula of Bridgman (2) which describes the normalized

21

distribution of sedimentation coefficients function. His statement provided for apparent, or idealized, distributions; we seek to show how the true distributions are obtained. Adequate adjustments are required for diffusion, and for concentration and pressure effects in modifying the form of the boundary gradient curve as it is displaced with time.

Both protein and polymer physical chemists now have used our extended procedures, often with appreciable success. In general, the proteins are dissolved in aqueous buffer media, where pressure-dependent effects are negligible. The polymer chemist, on the other hand, must use organic solvents, and by working at the Flory temperature, concentration-dependent effects are all but eliminated. Unfortunately for him, pressure-dependent effects are thus introduced and corrections for them are required as a prerequisite for a suitable analysis.

The procedures for obtaining a molecular weight distribution from the sedimentation equilibrium experiment may be described as "direct" or "indirect." Recognizing the fact that the different optical methods of observation and the several ways in which the experimental records may be analyzed yield different types of average molecular weight, M_n, M_w, M_z, etc., Lansing and Kraemer (3) established the principle of indirect determinations. Moments of the distribution curve are used as parameters in a distribution function of reasonable mathematical form, one which contains the same number of adjustable constants as there are reliable averages. It can be a quite satisfactory solution of the problem of describing the heterogeneity if there is an independent way of knowing that the molecular weight distribution function will have a particular mathematical form. These methods apply in the case of ideal or pseudo-ideal solutions, or in those situations where satisfactory corrections can be made to convert apparent average molecular weights to true quantities. (This conversion was indeed a subject of Chapter I.)

In the direct method, originally described by Rinde (1), the molecular weight distribution is constructed from the observed experimental record. The actual determination of the details of the distribution has since been attempted on a number of occasions; we refer here to Wales et al. (4), Fujita (5), Donnelly (6), Provencher (7), and Scholte (8) for further details.

In this chapter we will discuss the means by which a distribution of sedimentation coefficients is obtained and how it may be related to the distribution of molecular weight in one case, polystyrene, and how in addition to this distribution, a conclusion can be drawn about the macromolecular conformation in solution in another case, dextran. We note here that we shall be concerned with weight fraction distributions which denote the relative weight of the macromolecule in the system.

Theory

Sought first of all are detailed mathematical analyses of the changing outline of the boundary gradient curve with time as it moves down the cell in the transport experiment. In this endeavor, it is assumed, first, that the boundary spreading is due solely to the distribution of the limiting sedimentation coefficient, s_0, values, in other words that there is no broadening of the gradient curve with time due to diffusion and that there has been no sharpening of this curve due to either or both of the effects of varying concentration and pressure with position in the cell. Then, the attempt will be made to show to what extent graphical and analytical procedures may serve for the correction of diffusion and pressure (polystyrene) or diffusion and concentration (dextran) dependencies, to produce the limiting or idealized distribution.

As the subject has advanced, the nomenclature and symbolization has been anything but standardized. The objective here is to provide a practical outline.

The fundamental relationship is

$$\frac{dc}{ds} = \frac{dc}{dr}\frac{dr}{ds} = \frac{dc}{dr} r\omega^2 t \tag{II-1}$$

($\ln r/r_{2m} = s\omega^2 t$). The cell position, r_{2m}, is the square root of the second moment of the boundary gradient curve.

The normalized weight frequency function for the limiting sedimentation coefficients, s_0, is

$$g(s_0) = \frac{1}{c_0}\frac{dc_0}{ds_0} \tag{II-2}$$

where c_0 is the total concentration. To correct for the sector shape of the cell, $dc_0 = (r/r_0)^2 dc$, and remembering that in the idealized situation, $s = s_0$,

$$g(s_0) = \frac{1}{c_0}\left(\frac{r}{r_0}\right)^2 \omega^2 rt \frac{dc}{dr} \tag{II-3}$$

In the presence of diffusion, concentration, and pressure dependencies, insertion of the experimental quantities required by the right-hand side of Equation (II-3) does not lead to $g(s_0)$; it gives an apparent distribution function which may be symbolized as $g^*(S)$, with $S = (1/\omega^2 t) \ln r/r_{2m}$. S defines a position in the cell, rather than a sedimentation coefficient for any particular dissolved species.

DIFFUSION

The correction for diffusion depends upon the fact that the diffusional boundary gradient curve spreading varies with the square root of the time $(t^{1/2})$ while the contribution due to heterogeneity of sedimentation coefficients is proportional to the first power of time (9). Thus, under suitable experimental conditions, an extrapolation to infinite time should eliminate the effect of diffusional boundary spreading. For any given experiment the $g*(S)$ curve is computed for different times, then this curve is extrapolated to infinite time. Theoretical work by Gosting (9a) has defined the conditions under which the extrapolation is proper. One seeks to find a region of time in which $g*(S)$ is a linear function of $(t)^{-1}$. Unfortunately, this linear region is not often reached (see Appendix B), and extreme care must be taken in the interpretation of the data.

PRESSURE

Relative to the correction of the $g*(S)$ curve for pressure, there remain unsolved problems. However, Fujita (10) has made a study of the effects of a pressure dependence on the sedimentation coefficient which leads to a quite good, if somewhat provisional, answer. With pressure dependence, the apparent sedimentation coefficient is a function of distance from the meniscus, and there is a continuous concentration gradient from the boundary region to the bottom of the cell. Thus, there is no longer a "plateau region."

The effect originates in variations with pressure of the molar friction coefficient, f, and the partial specific volume, $\bar{v}$, of the solute, and in the density of the solvent system, ρ. These variations may be assumed to follow the following linear relationships; they are written for the case in which s depends on pressure alone.

$$f = f_0(1 + \lambda P) \tag{II-4}$$

$$\bar{v} = \bar{v}_0\left(1 - \frac{1}{\kappa}P\right) \tag{II-5}$$

$$\rho = \rho_0(1 + \beta P) \tag{II-6}$$

The overall variation of sedimentation coefficient is also assumed to follow a linear law:

$$s = s_0[1 - kP] \tag{II-7}$$

where

$$s_0 = \frac{M(1 - \bar{v}_0\rho_0)}{f_0} \quad (= \text{value of } s \text{ at } P = 0)$$

and

$$k = k(\lambda, \beta, \kappa) = \lambda + \frac{\bar{v}_0\rho_0(\beta - 1/\kappa)}{1 - \rho_0\bar{v}_0}$$

The molar friction coefficient which corresponds to s_0 is designated as f_0, and P is the pressure measured from one atmosphere.

Fujita has solved the continuity equation [Equation (A-5)] for the case of a single solute with a linear dependence of s with pressure when D, the diffusion coefficient, remains constant.

The variation of pressure, P, with radial distance, r, is

$$P = \tfrac{1}{2}\omega^2 r_0^2 \rho_0(y - 1) \tag{II-8}$$

$$y = (r/r_0)^2 \tag{II-8a}$$

So, the variation of s with r becomes

$$s = s_0[1 - m(y - 1)] \tag{II-9}$$

with

$$m = \tfrac{1}{2}k\omega^2 r_0^2 \rho_0 \tag{II-9a}$$

The radial dilution law is now modified to read

$$\frac{c_0}{c} = \left(\frac{r}{r_0}\right)^2 [1 - m(y - 1)]$$

where c is the concentration which corresponds to r.
Then,

$$dr/ds = r\omega^2 t[1 - m(y - 1)]$$

and

$$g(s_0) = \frac{r^3 \omega^2 t[1 - m(y - 1)]^2}{r_0^2 c_0} \frac{dc}{dr} \tag{II-10}$$

In this final expression for $g(s_0)$, it has been assumed that there is no broadening of the boundary gradient curve due to diffusion, nor is there any sharpening due to concentration-dependent effects. The expression contains two factors $[1 - m(y - 1)]$, one from dr/ds and the other from the radial dilution law.

Experimentally the value of the parameter m is obtained from the statement

$$\frac{d \ln (r/r_0)}{d(\omega^2 t)} = s \text{ (an "apparent" value)} = s_0[1 - m(y - 1)] \tag{II-9b}$$

The plot of s versus the quantity $(y - 1)$ is linear. Then, as $(y - 1) \rightarrow 0$, the value of s_0 is obtained, while the slope of the line leads to a value of the parameter m.

CONCENTRATION

In general, both s and D vary not only with the concentration of a given solute but also with concentrations of other solutes which may be present in the solution. Thus, to apply the idea that the concentration gradient curves, actually refractive index gradient curves, obtained in sedimentation velocity experiments with polydisperse solutes may be transformed to a distribution function, it is necessary to extrapolate results obtained at finite concentrations to infinite dilution where effects of concentration and coupling of solute flows disappear. The latter effects are probably small enough to be neglected in most applications. However, in nonideal solutions, the concentration-dependent effects must be eliminated in one way or another to obtain a distribution with physical significance. To this end the polymer chemist may operate at the Flory temperature, but such an option is presumably not available to the protein chemist.

Unfortunately, no theory is as yet available to guide the extrapolation. The difficulty arises because, since s depends upon the concentrations of all the solutes present, a complex Johnston–Ogston effect is involved, and the mathematical problems introduced have been insurmountable, to date. Various empirical procedures have been proposed (*11–13*) to provide a provisional result. Their use has been illustrated at length, both in the original articles and in the Fujita monograph (*14*). Comparisons of the results of the several procedures show that all three of them give substantially the same limiting distribution. The methods of Baldwin and of Gralén and Lagermalm are not greatly different; their use is to be preferred, especially when data are not available at very low concentrations, as is usually the case.

CORRELATION OF MEASUREMENTS OF $g(s_0)$ AND $f(M)$. HOMOLOGOUS POLYMERS

One finds in the polymer chemistry literature empirical relationships between limiting viscosity number, diffusion coefficient, and sedimentation coefficient with molecular weight. For the sedimentation coefficient,

$$s = KM^\alpha \tag{II-11}$$

in which the values of the constants K and α depend upon the nature of the solvent, the conformation of the polymer molecule in it, and the extent to which the latter is permeated by this solvent. Much theoretical work has shown that in the special case of the Flory temperature, the value of α should be $\frac{1}{2}$.

For a continuous distribution,

$$g(s_0)\, ds_0 = f(M)\, dM \tag{II-12}$$

or

$$f(M) = g(s_0)\, ds_0/dM$$
$$= g(s_0)\alpha K(s_0/K)^{\alpha - 1/\alpha}$$

The quantities $f(M)$ and $g(s_0)$ are the differential distribution functions of M and s_0. The subscripts zero now have their usual significance; the apparent s data have been corrected to eliminate diffusion effects, and pressure and concentration dependencies.

Again, where s_0' corresponds to M',

$$\int_0^{s_0'} g(s_0) \, ds_0 = \int_0^{M'} f(M) \, dM$$

$$G(s_0') = F(M') \tag{II-13}$$

These are the integral distribution functions which correspond. Thus, it is required to have numerical values for α and K to make the transformation to obtain a molecular weight distribution of a sample from an investigation of its distribution of s in the velocity ultracentrifuge. This operation is to be discussed in connection with data for a sample of polystyrene which have been taken in the transport experiment under the pseudo-ideal condition (Flory temperature) to eliminate concentration-dependent effects.

The intent can be inverted. If the broader details of the molecular weight distribution curve are made available by equilibrium ultracentrifugation and they are combined with the sedimentation coefficient distribution, it becomes possible to find the relation between sedimentation coefficient and molecular weight. This is illustrated in the next section by the results from such a study with a fractionated dextran in solution.

Experiment I. Polystyrene–Cyclohexane at 34.2°C

Observations of boundary spreading in sedimentation velocity experiments (15) have been used to indicate the route by which the solute molecular weight distribution curve may be obtained. Records of the sedimentation behavior and properties of the system polystyrene–cyclohexane at the Flory temperature have been adapted to this end. The polymer was a Dow Chemical Company material, designated as 19F.

In the interpretation of the data, the constants α and K, Equation (II-11), were adjusted from data and other information already in the literature and in such a way that the number average molecular weight for the polymer would correspond to 200,000, a value determined by Grandine (16) by using osmotic pressure determinations. The literature data are those of McCormick (17) and Cantow (18), in order

$$s = (1.69 \times 10^{-2})M^{0.48}$$

$$s = (1.35 \times 10^{-2})M^{0.51}$$

In making the transformation from $g(s_0)$ versus s_0 to $f(M)$ versus M, the parameters $\alpha = 0.50$ and $K = 1.47 \times 10^{-2}$ were selected. Furthermore, it was assumed that no correction for diffusional spreading of the boundary gradient curve was necessary.

An observed boundary gradient curve for the polystyrene 19F-cyclohexane system at 34.2°C, the Flory temperature, is shown in Figure II-1. In this

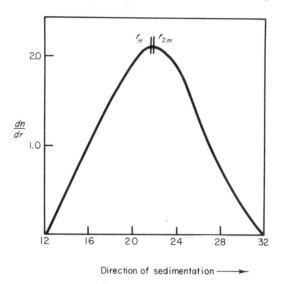

Direction of sedimentation ⟶

Figure II-1. Typical boundary gradient curve for polystyrene 19F in cyclohexane at 34.2°C. Coordinates are in arbitrary units. The radial positions are those of the maximum height, r_H, and for second moment, r_{2m}, of the boundary gradient curve. Speed, 59,780 rpm. Redrawn from Blair and Williams (*15*). Reprinted with modification from *J. Phys. Chem.* **68**, 161 (1964). Copyright (1964) by the American Chemical Society. Reprinted by permission of the copyright owner.

figure, positions both of the maximum height (r_H) and of radial position (r_{2m}) of the curve are shown. In carrying out the subsequent computations for the sedimentation coefficients the radial distance, r_{2m}, was used.

The value of the pressure-dependent parameter m, Equation (II-9a), was determined from a plot of $\ln (r_{2m}/r_0)$ versus $\omega^2(t - t_0)$ where $t - t_0$ is a corrected time. An additional plot of m/r_0^2 versus the square of the rotational speed is linear, to give confidence that the value of m is a significant one.

The data of Figure II-1, representing a sedimentation velocity experiment (polystyrene 19F at concentration 0.504 g/100 ml, temperature 34.2°C, and rotor speed 59,780 rpm), dn/dr versus r, were transformed to $g(s_0)$ and s by Equation (II-10) and

$$s = \frac{\ln (r/r_0)}{\omega^2 t}$$

respectively, to give the distribution curve of limiting sedimentation coefficients. From it the molecular weight distribution, $f(M)$ versus M, was achieved, with $\alpha = 0.50$ and $K = 1.47 \times 10^{-2}$. The resulting $f(M)$ versus M curves for the two cases, with and without pressure dependence correction of sedimentation coefficient, appear in Figure II-2. The curve which represents the most

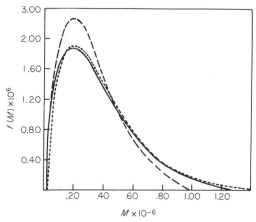

Figure II-2. Molecular weight distribution curve for polystyrene 19F system: — — — —, not corrected for pressure ($a = 0.50$, $k = 0.0147$); ———, corrected for pressure dependence ($a = 0.50$, $k = 0.0147$); - - - - - -, most probable distribution for $M_n = 200,000$. Redrawn from Blair and Williams (15). Reprinted with modification from *J. Phys. Chem.* **68**, 161 (1964). Copyright (1964) by the American Chemical Society. Reprinted by permission of the copyright owner.

probable distribution, with $M_n = 200,000$, an extremely good representation of the experimental data, is also shown.

In order to verify the general validity of this curve, numerical integrations were performed to obtain the number and weight average molecular weights. By definition

$$M_n = \frac{\int f(M)\,dM}{\int \frac{1}{M} f(M)\,dM} \qquad \text{(II-14)}$$

and

$$M_w = \frac{\int M f(M)\,dM}{\int f(M)\,dM} \qquad \text{(II-15)}$$

The numerical results are found to be $M_n = 207,000$ and $M_w = 375,000$.

Had we elected to accept the McCormick values for α and K the figures for M_n and M_w would have been 243,000 and 488,000, respectively, values which are almost certainly too high. The true value for M_w is probably in the neighborhood of 425,000, with $M_w/M_n \cong 2$. The proper evaluation of α and K is an extremely difficult matter; the final result for the average molecular weights is highly sensitive to the values selected for both parameters.

Experiment II. Dextran–Water at 25°C

The efficacy of any plasma extender substance depends, among other things, upon its molecular size and conformation. The solutions of gelatin, polyvinylpyrrolidone, and dextran commonly used systems for this purpose are characterized by solute polydispersity. There appears to be some relationship between molecular size and retention time in the system and, even though it may be only one criterion, we have worked out methods to describe the polydispersity of their solutions by ultracentrifugal analyses. Their development and application to a representative dextran fraction is illustrated here.

For the sedimentation equilibrium experiment, a practical means for the description of solute heterogeneity is found in a comparison of the several thermodynamic average weights over the cell, M_n, M_w and M_z. These data are obtained by the conversion of apparent values to the corresponding thermodynamic quantities. As a general statement, it may be noted that the greater the spread of these quantities, the greater is the heterogeneity; on the other hand, if they all have the same value, the solute must be homogeneous.

The method employed is to fit an assumed parametric function to these observed average molecular weights to obtain a distribution function for molecular weights. Such an analysis is not at all sensitive to minor details in the distribution curve, and it has been found useful to combine this result with the distribution of sedimentation coefficients from an investigation of the same sample by velocity ultracentrifugation. It then becomes possible to find the relation between sedimentation coefficient and molecular weight, and through it, some information about the molecular form of the solute.

Sedimentation Equilibrium

For application to solutions of plasma extender materials at sedimentation equilibrium, the Lansing-Kraemer (3) logarithmic number distribution to obtain $f(M)$ is reasonably satisfactory. It is described by the formula

$$f(M)\, dM = dW = \frac{W}{M_n \beta (\pi)^{1/2}} \exp\left(-\frac{1}{\beta} \ln \frac{M}{M_0}\right)^2 dM \qquad \text{(II-16)}$$

where the notation is as follows:

dW = weight of material having molecular weight between M and $M + dM$

W = total weight of material, normalized to unity

M_0 = molecular weight at the maximum value of dW/dM

M_n = number average molecular weight

β = nonuniformity coefficient

The nonuniformity coefficient is obtained from a combination, either of M_n or M_w, with M_p, with the latter pair being more readily accessible from the experimental data. The quantity M_p is the most probable molecular weight.

Characteristic data for a representative dextran fraction (Commercial Solvents) have been found to be $M_0 = 33,000$, $M_n = 42,000$, $M_w = 65,000$, with $\beta = 0.94$. The actual molecular weight differential distribution curve for dextran (*11*) appears as Figure II-3, where

$$f(M) = \frac{dW}{dM} = \frac{1}{42,000 \times 0.94 \times 1.77} \exp\left(-\frac{1}{0.94}\ln\frac{M}{33,000}\right)^2$$

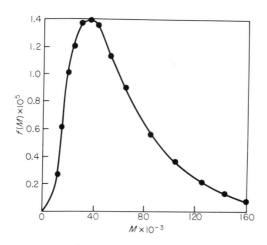

Figure II-3. Differential molecular weight distribution curve for a typical plasma extender dextran fraction. Redrawn from Williams and Saunders (*11*). Reprinted with modification from *J. Phys. Chem.* **58**, 854 (1954). Copyright (1954) by the American Chemical Society. Reprinted by permission of the copyright owner.

SEDIMENTATION TRANSPORT

The dextrans in solution form nonideal systems, so that any observed distribution of sedimentation coefficients is concentration-dependent. Thus, an extrapolation procedure must be applied to obtain the limiting distribution, $g(s_0)$. It is proper to assume that pressure-dependent effects are absent in the aqueous system. Then, in the limit of zero concentration, with $D = 0$,

$$g(s_0) = \frac{dn}{dr} \frac{\omega^2 rt}{n_1 - n_0} \tag{II-17}$$

where dn/dr is the refractive index gradient, and n_1 and n_0 are the refractive indexes of solution and solvent, respectively. This expression gives the relative amount of solute with sedimentation coefficient s_0.

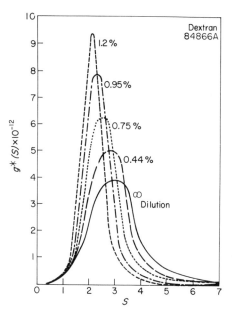

Figure II-4. $g*(S)$ versus S curves at several finite concentrations and the extrapolated curve for infinite dilution for the dextran fraction. Redrawn from Williams and Saunders (*11*). Reprinted with modification from *J. Phys. Chem.* **58**, 854 (1954). Copyright (1954) by the American Chemical Society. Reprinted by permission of the copyright owner.

The apparent distribution function for s at any concentration, $g*(S)$, is calculated from the experimental records by means of the formula

$$g*(S) = r \frac{dn}{dr} \frac{\omega^2 t}{A_{t \to 0}} \exp\left(2s\omega^2 t\right) \tag{II-18}$$

where the notation is as follows:

$\quad S =$ position in the cell corresponding to $(1/\omega^2 t) \ln r/r_{2m}$

$\quad t =$ time in seconds, following the attainment of the steady-state velocity

r = radial position in the cell for material of sedimentation coefficient s, at time t

dn/dr = height of boundary gradient curve at r

A = area under the boundary gradient curve, corrected to zero
time

$\exp(2s\omega^2 t)$ = radial dilution factor

Then by plotting the whole $g^*(S)$ versus $1/rt$ curves and extrapolating to infinite time, the effects of diffusion in broadening the curve are eliminated.

In Figure II-4 are presented $g^*(S)$ versus S curves, after correction for diffusional spreading, for the selected sample at four finite concentrations, and following extrapolation of these curves, the $g(s_0)$ versus s_0 curve, corresponding to infinite dilution.

COMBINATION OF VELOCITY AND EQUILIBRIUM DISTRIBUTIONS

Having determined the functions $g(s_0)$ and $f(M)$, they may be combined in the following manner, using Equations (II-12) and (II-13). A double plot of integral distribution curves, $F(M)$ versus M and $G(s_0)$ versus s_0, for the

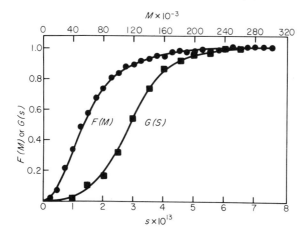

Figure II-5. Double plot of the integral distribution functions, $F(M)$ and $G(s)$, as functions of M and s_0, respectively, for the representative dextran. Redrawn from Williams and Saunders (*11*). Reprinted with modification from *J. Phys. Chem.* **58**, 854 (1954). Copyright (1954) by the American Chemical Society. Reprinted by permission of the copyright owner.

dextran fraction is given in Figure II-5. If from this plot the values of the two abscissas at equal ordinates are taken, the s_0' value which corresponds to a given M' is determined (*19*). In this way, the entire curve of s_0 versus M can

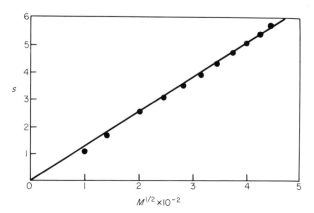

Figure II-6.   Sedimentation coefficient versus $M^{1/2}$ for the dextran fraction. Redrawn from Williams and Saunders (*11*). Reprinted with modification from *J. Phys. Chem.* **58,** 854 (1954). Copyright (1954) by the American Chemical Society. Reprinted by permission of the copyright owner.

be constructed. In constructing the plot, Figure II-6, the abscissa parameter chosen was $M^{1/2}$. It is seen at once that to a good approximation s_0 varies linearly with $M^{1/2}$. The exponent $\frac{1}{2}$ is characteristic of flexible, long-chain molecules in which the shape of the chain is described by the random flight model—the random coil.

By similar experimental procedures it has been demonstrated that in plasma-extender gelatin solutions the s_0 to $M^{1/2}$ relationship also obtains; the solute has been denatured in the process of manufacture.

Summary

In a polydisperse system the sedimentation velocity experiment may provide useful information about solute heterogeneity. The general approach has had its successes, with both internal and external evidence that the distributions of s_0 (and M) obtained are real.

Actually, the procedures described are representative of an earlier period in their application; they are perhaps now somewhat naive. The present state of theory is well illustrated by a recent article of Oncley (*20*) entitled "Evaluation of Lipoprotein Size-Density Distributions from Sedimentation Coefficient Distributions Obtained at Several Solvent Densities." A more recent practical application is found in a work of Wales and Rehfeld (*21*), "Molecular Weight Distribution by Velocity Ultracentrifugation," which describes a general procedure for determining molecular weight distributions in linear, neutral polymers. Information about boundary spreading, the usual ancillary data, and a partial knowledge of the dependence of intrinsic viscosity on molecular weight are required. The report contains convincing verification of the results by an independent method.

References

1. H. Rinde, Dissertation, Uppsala (1928).
2. W. B. Bridgman, *J. Amer. Chem. Soc.* **64**, 2349 (1942).
3. W. D. Lansing and E. O. Kraemer, *J. Amer. Chem. Soc.* **57**, 1369 (1935).
4. M. Wales, J. W. Williams, J. O. Thompson and R. H. Ewart, *J. Phys. Chem.* **52**, 983 (1948).
5. H. Fujita, *J. Chem. Phys.* **32**, 1739 (1960).
6. T. H. Donnelly, *J. Phys. Chem.* **70**, 1862 (1966).
7. S. W. Provencher, *J. Chem. Phys.* **46**, 3229 (1967).
8. T. G. Scholte, *Ann. N.Y. Acad. Sci.* **164**, 156 (1969).
9. R. L. Baldwin and J. W. Williams, *J. Amer. Chem. Soc.* **72**, 4325 (1950); J. W. Williams, R. L. Baldwin, W. M. Saunders and P. G. Squire, *ibid.* **74**, 1542 (1952).
9a. L. J. Gosting, *J. Amer. Chem. Soc.* **74**, 1548 (1952).
10. H. Fujita, *J. Amer. Chem. Soc.* **78**, 3598 (1956).
11. J. W. Williams and W. M. Saunders, *J. Phys. Chem.* **58**, 854 (1954).
12. R. L. Baldwin, L. J. Gosting, J. W. Williams and R. A. Alberty, *Discuss. Faraday Soc.* **20**, 13 (1955).
13. N. Gralén and G. Lagermalm, *J. Phys. Chem.* **56**, 514 (1952).
14. H. Fujita, "The Mathematical Theory of Sedimentation Analysis." Academic Press, New York, 1962.
15. J. E. Blair and J. W. Williams, *J. Phys. Chem.* **68**, 161 (1964).
16. L. D. Grandine, Jr., Dissertation, University of Wisconsin (1952).
17. H. W. McCormick, *J. Polym. Sci.* **36**, 341 (1959).
18. H. J. Cantow, *Makromol. Chem.* **30**, 169 (1959).
19. J. W. Williams, *J. Polym. Sci.* **12**, 351 (1954).
20. J. L. Oncley, *Biopolymers* **7**, 119 (1969).
21. M. Wales and S. J. Rehfeld, *J. Polym. Sci.* **62**, 179 (1962).

PART **2**

SELF-ASSOCIATION REACTIONS
IN PROTEIN SYSTEMS

CHAPTER **III**

Simultaneous Sedimentation and Chemical Equilibrium in β-Lactoglobulin B, Chymotrypsinogen A, and Lysozyme Solutions

In addition to the primary, secondary and ternary structures, there is often another kind of organizational order found in the proteins. The existence of this fourth type of structure was first revealed by the early ultracentrifuge studies of Svedberg and his associates, who showed that there are situations in which molecular weight can be substantially modified by relatively small changes in environment—pH, ionic strength, and temperature of the solution—but always in an orderly manner. The process has become known as self-association, a reaction which involves in its inception a monomeric unit.

More recently, several transport methods, for example, sedimentation velocity, free electrophoresis, countercurrent distribution, and analytical gel chromatography, have been employed in the study of the chemically reacting protein systems. However, there have remained some uncertainties in the interpretation of the data, and it has seemed expedient to turn to the sedimentation equilibrium experiment in the further investigation of the association mechanisms. Extended mathematical analyses may be said to have begun with Adams and Fujita (*1*), enhanced in the sense that allowance for effects of solution nonideality is included. Tiselius (*2*) had very early given the first mathematical treatment of the problem, but it was restrictive because of ideal solution behavior being assumed for each protein species. The same situation obtained in the earlier derivations that enabled the application of the osmotic pressure and light-scattering techniques to the problem. The status of the subject in 1963, that is, prior to the appearance of the Adams-Fujita report, is set down in an excellent recent survey by Nichol *et al.* (*3*).

Since 1963 there has been greatly renewed interest in both theory and experiment for protein self-associations. A description is given in this chapter of self-association reactions which involve β-lactoglobulin B, lysozyme, and chymotrypsinogen A in well-defined solvent media, and as studied by us. For the first of these proteins, the data may be explained by a monomer–dimer stoichiometry, for the second the data are about equally well accounted for with either monomer–dimer reaction or random association behavior, while for the chymotrypsinogen A the reaction appears to be one of random association. But even in some of those cases where there is uncertainty in the selection of the reaction type (high or low values of association constant) it will evolve that reasonably good thermodynamic data may be obtained.

The sedimentation equilibrium approach has been selected for the study because with proper experimental conditions it may provide more than one of the common average molecular weights which characterize the association system at the several radial distances in the solution cell. The protein must meet the requirement of purity; in addition, the chemical equilibrium must be established rapidly. Our selection of the three proteins has been made with these criteria in mind, and also because probable courses of the reactions are more readily assigned if only one association equilibrium constant is adequate for the analysis of the data, as was the case for these proteins. For the β-lactoglobulin B system at the lower temperatures the effects of solution non-ideality became more pronounced in the regions of higher solute concentration and an additional virial coefficient was employed; otherwise a single such coefficient, together with an association constant, sufficed to account for the form of the concentration versus distance, c versus r curve, found directly by the experiment, or for the $M_1/M^a_{w(c)}$ versus c curve as transformed from it.

In the development of the mathematical apparatus to provide for simple and practical equations, certain reasonable assumptions are made throughout. They are as follows:

(a) All species, monomer, dimer ... (species 1, 2, ...), have the same partial specific volume and refractive index increment.

(b) For very dilute solutions, we may expand $\ln y_1$, $\ln y_2$, etc., in powers of c, according to the scheme

$$\ln y_1 = M_1 B_1 c + \cdots = M_1 B c$$

$$\ln y_2 = M_2 B_2 c + \cdots = 2 M_1 B c + \cdots$$

$$\ln y_i = i M_1 B c$$

As an approximation, we have set $B_1 = B_2 = \cdots = B$ and have neglected the higher terms in concentration. While this assumption may appear to be arbitrary, it is not unreasonable. It does have great advantage in the reduction of a complicated problem in those situations where the data cannot be explained simply by an association reaction, ideal solution behavior, in the sense that the second virial coefficient is zero.

In the interpretation of the data use is made of the basic equation

$$\frac{1}{M^a_{w(c)}} = \frac{1}{M_{w(c)}} + Bc + 0(c^2) + \cdots \tag{III-1}$$

It is superficially like one which has already been used for solute systems with continuous (rather than discrete) distributions of molecular weight. Proof of the validity of Equation (III-1) in a two-component system when applied to the dimer association of solute species—still a single component—is to be found in the Adams-Fujita (*1*) article. (Complete generalization is apparently a subject for further consideration.) It teaches that $M^a_{w(c)}$, the apparent weight average molecular weight which corresponds to the protein component concentration c, is a unique function of this total solute concentration. Operationally, this means that the $M^a_{w(c)}$ versus c data, when plotted from a series of individual experiments at different initial solute concentrations and at different rotor speeds and solution depths, must superimpose to form a single continuous curve of $M^a_{w(c)}$ versus c. The quantity $M^a_{w(c)}$ is defined by the relation

$$M^a_{w(c)} = \frac{1}{rc_{(r)}} \frac{dc_{(r)}}{dr} \frac{RT}{(1 - \bar{v}\rho)\omega^2} \tag{III-2}$$

It is not to be confused with the weight average molecular weight over the whole cell; it is a quantity which corresponds to a fixed protein concentration, which is the same as noting that it is descriptive of the molecular weight average at a fixed radial position r. The density ρ refers to that of the dialyzed solvent; the partial specific volumes of all species are considered to be constant, $\bar{v}$.

The data are obtained from experiments performed at a number of initial protein concentrations and at different low rotor speeds with constant solution depth. The molecular weight averages, $M^a_{w(c)}$, at each of several radial distances in the cell may be regarded as being equivalent to the information which is made available by a series of light-scattering experiments at the corresponding concentrations. The basic concentration versus distance in the cell curves are transformed and collected to give a single $M^a_{w(c)}$ versus c curve. (Such behavior may not be observed if the protein contains an even quite small amount of impurity.) This curve is characterized by an initial increase in $M^a_{w(c)}$ corresponding to low concentrations of the protein component, indicative of protein association. Then, depending upon the reaction type, such curves may become concave downward or concave upward with increasing concentration. When concave downward, the presumption is that a simple kind of reaction is involved, one of the monomer–dimer type. (Tests with simulated data demonstrate that even the curves for a monomer–n-mer system, with $n > 2$, yield sigmoidal curves which are both concave downward and concave upward.) This curve may then level off or it may actually decrease, following the initial increase. Such a subsequent decrease is an indication of a relatively strong solution nonideality which becomes the predominating influence at the higher protein concentrations. The advantage of conducting the investigation over as large as possible a concentration interval, as opposed to attempting to extract information from a few, even two, individual experiments, cannot be overemphasized.

Perhaps it should be noted that both equilibrium constant and second virial coefficient are really measures of solution nonideality, for all intermolecular interactions which occur in the system may be described in terms of the excess chemical potentials, $RT \ln y_i$. We elect to sort out those which go with actual protein self-association, attempt to assign a type or mechanism to the reaction, and evaluate them as the equilibrium constant. The remaining interactions might have been earlier described by the word "solvation." However, assignment of the exact nature of these interactions is impossible for the present; their effect is measured by the other thermodynamic quantity, namely, the second (third, etc.) virial coefficient.

To calculate equilibrium constants and nonideality terms, the $M^a_{w(c)}$ versus c curve may be transformed into one of the reciprocal apparent molecular weight function, $M_1/M^a_{w(c)}$, versus c. So, the new curve is represented by the equation

$$M_1/M^a_{w(c)} = M_1/M_{w(c)} + BM_1c \qquad \text{(III-1a)}$$

which is descriptive of the association reaction in a nonideal solution and at sedimentation equilibrium. Another reciprocal apparent molecular weight function is sometimes utilized for this purpose, namely

$$M_1/M^a_{n(c)} = M_1/M_{n(c)} + \tfrac{1}{2}BM_1c + \cdots \qquad \text{(III-1b)}$$

where M^a_n, the apparent number average molecular weight, is computed from the identity

$$M^a_{n(c)} = \frac{c}{\displaystyle\int_0^c dc'/M^a_{w(c')}}$$

How the information provided by the $M_1/M^a_{w(c)}$ versus c curve or that of $M_1/M^a_{n(c)}$ versus c curve is utilized is discussed later. For the present we note only that it is necessary to know the molecular weight of the monomer, M_1. This datum is not always readily attainable. Ideally it should correspond to the component monomer of Casassa and Eisenberg (4), but in systems with chemical reactions taking place, extrapolations of data to vanishingly low protein concentrations are often unreliable. Alternatives are to sum the amino-acid residue weights, or to measure the apparent molecular weight at several relatively low protein concentrations under conditions of pH, ionic strength, and temperature such that association tendencies are largely suppressed, followed by suitable extrapolations. With data for M_1 the analysis may depend primarily on the application of extensions of equations originally presented by Steiner (5) to include the effects of solution nonideality. In essence, one studies how an observed average molecular weight ($M_{w(c)}^a$ or $M_{n(c)}^a$) should depend upon the total solute concentration for the type of reaction being assumed and then compares it with the experimental record to determine the propriety of the model. With a definitive model, the corresponding equilibrium constant is then amenable to computation.

An alternative method, one which we have not used, is founded upon a direct conversion of the total equilibrium concentration at any radial distance into its constituent species concentrations. The expression for the concentration as it changes with radial position in the cell is set down as the sum of exponential terms which contain unknowns to be determined in both amplitude and exponent of each term. For the self-associating system the exponents are generally known, thus the solutions of the equations are correspondingly simplified. Much of a large amount of relevant theory which has recently made its appearance in the literature is written in these terms, and with the Rayleigh optical system for the ultracentrifuge it ought to become a preferred method.

As already mentioned, we present here data for three proteins which undergo self-association, when solution conditions are properly chosen. In the first two instances literature data had indicated reactions of the monomer–dimer type; our experiments were conducted as a test of the efficiency of the mathematical analyses when applied to sedimentation equilibrium data of what we believe to be as high precision as is presently accessible. The systems are as follows:

    (1) β-Lactoglobulin B at pH 2.64, $I = 0.16$, and several temperatures (6, 6a)
    (2) Lysozyme at pH 7.0, $I = 0.20$, and $T = 25°C$ (7, 7a)
    (3) Chymotrypsinogen A at pH 7.9, $I = 0.03$, and $T = 25°C$ (8)

The meniscus depletion (or high speed) sedimentation equilibrium method was not used. The theories for the combined sedimentation and chemical equilibrium which will be given do not take into account any pressure dependence of the equilibrium constants for the reactions. Pressure gradients in the cell, through their influence on these constants, may modify the redistribution of the solute species along the cell column; their influence is kept small by operation at low values of a quantity λ, Equation (A-17) (i.e., $\lambda M_w \sim 1$). For safe operation, the values of $M_{w(c)}^a$ at points of equivalent concentration must be invariant with λ, a quantity which is a function of both rotor speed and solution column depth.

Relative to the experimental data themselves we shall remark only that all known precautions were taken with the preparation of the proteins themselves and their buffered solutions, and with the ancillary measurements of specific refraction increments, dn/dc, the apparent specific volume, $\bar{v}$, etc. As indicated, all ultracentrifuge experiments were of the low-speed, short solution column variety. The optical system was the standard Rayleigh/schlieren arrangement. The quantity $M_{w(c)}^a$ as a function of c was calculated by operations for which detailed descriptions can be found in recent publications (6–8).

There are a number of procedures for analyzing the data to provide information about the reaction mechanism (model) and to evaluate the several thermodynamic characteristic constants. For our treatments of the data for the three protein systems we have utilized, first of all, two schemes of Adams and associates—a successive approximation (but one to which we have added means of fitting the data to the model with simultaneous minimization of residuals) (9) and the approach of Chun et al. (10) based upon the recognition that if apparent molecular weight averages and other apparent functions are properly combined, the nonideality terms can be made to drop out. In addition, and especially for those cases in which the monomer–dimer model is appropriate, we have used two closed-form methods derived from the statement, Equation 22, of Adams and Fujita (1), in which the apparent weight average molecular weight is expressed as an explicit function of concentration, as follows:

$$2(M_1/M_{w(c)}^a) - 1 = (1 + 4K_2c)^{-1/2} + 2BM_1c \qquad \text{(III-3)}$$

As each of the three protein systems is individually treated more definitive statements will be made about the manner in which a model has been assigned and by which the association constant, K_2, and the virial coefficients have been given numerical values.

β-Lactoglobulin B

It has been reported, based largely on the results of light-scattering measurements, that in acid solution β-lactoglobulin B undergoes a rapid dimerization. Our sedimentation equilibrium experiments were designed to test extended methods to study reversibly associating systems, the development of which in effect began with Adams and Fujita (1). β-Lactoglobulin B served the purpose of test substance; it forms solutions in which monomer and dimer predominate and, in a solvent which under proper conditions can be treated as a single component, gives essentially a two-component system (4).

In addition to testing the adequacy of several of the theoretical formulations for the evaluation of the reaction association constant, K_2, and the virial coefficient (or coefficients), we have sought to provide a broader thermodynamic understanding of protein self-association reactions by conducting the experiments at several fixed temperatures. An extensive series of experiments was carried out with the β-lactoglobulin B system at pH 2.64, $I = 0.16$ at five temperatures from 5° to 35.5°C (6). Previous investigations by Timasheff and Townend (11) and by Albright and Williams (6a) had indicated that under substantially the same solution conditions the predominant reaction is one of the monomer–dimer type. However, in spite of the simplicity of the model, there are certain aspects which require great care in the analysis and an indication of the problems which attend the evaluation of the several thermodynamic parameters is of considerable interest.

The actual data were obtained by the low-speed, short column of constant depth, sedimentation equilibrium experiment, with interference optics used to register the final redistribution of the solute component of the two species. Representative data taken at 15° and 25°C are shown in Figure. III-1. The con-

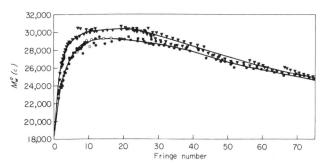

Figure III-1. Plot of apparent weight average molecular weight as a function of concentration, expressed in fringes, for β-lactoglobulin B in solution at pH 2.64, $I = 0.16$. Upper curve, $T = 15°C$; lower curve, $T = 25°C$. Concentration determined by using $\varepsilon_{1\ 78}^{\%} = 9.2$ dl/g-cm. Reproduced from Visser et al. (6). Reprinted from *Biochemistry* **11**,2634 (1972). Copyright (1972) by the American Chemical Society. Reprinted by permission of the copyright owner.

centration scale is expressed as fringe number. The solid lines in each case are interpolation curves which were used to obtain the quantities required for the analyses. The several kinds of points which provide the information for the interpolation curves, open and closed circles and triangles, represent data derived from a single experiment at a given initial solute concentration. The average deviation in $M^a_{w(c)}$ was about 350–400 g/mole.

The data taken from a number of experiments at different initial concentrations and at the final temperatures form a continuous curve within the expected experimental precision. The rotor speeds were varied from 6,160 to 16,200 rpm, resulting in a maximum of 20 fringes across the 12 mm cell. The single unique curve descriptive of the data at a given temperature indicates that the chemical reaction is reversible, that pressure effects are of negligible influence over the cell, and that the protein is of good quality.

In these experiments and in the earlier ones of Albright and Williams (6a) the protein concentrations were determined spectrophotometrically. For the extinction coefficient, the value $\varepsilon_{278}^{1\%} = 9.6$ dl/g-cm was at first utilized; it was taken from determinations reported by Townend et al. (12). Using this value, the Visser et al. (6) results are in better accord with the light-scattering molecular weight data of Timasheff and Townend (11) than they are with those of Albright and Williams from sedimentation equilibrium. In the attempt to discover the reason for the discrepancies, differential refractometry was employed in addition to spectrophotometry for the determination of the protein concentration. It was found that agreement between light-scattering and sedimentation equilibrium methods could be obtained only if the extinction coefficient was taken to be 9.1, incidentally a value determined later and independently for us by Dr. C. N. Pace (13) at Texas A. & M. University. We have sought to make allowance for residual light-scattering by adopting the value 9.2. It appears that if close agreement in the apparent weight average molecular data from different investigators is to be obtained by using spectrophotometry, perturbations due to light scattering from the protein solutions must be carefully avoided either by millipore filtration of solution or by control.* It is our opinion that differential refractometry is a better procedure for the purpose.

TREATMENT OF DATA. MONOMER–DIMER MODEL

To obtain the characteristic constants, K_2, B_1M_1 ($= BM_1$ when one virial coefficient is used), and if necessary, B_2M_1, for the β-lactoglobulin B systems three types of analysis have been applied.

(a) The Van Holde-Rossetti (14) method for an indefinite (open, isodesmic) reaction has been modified by Deonier and Williams (7) to be applicable to the monomer–dimer case. The working equation then takes the form

$$\frac{R_a^2}{[2(1 - BM_1R_ac) - R_a]^2} - 1 = 4K_2c \qquad \text{(III-4)}$$

---

* It may be of interest at this point to insert a remark about the magnitude of the error introduced by an error of $\sim 4\%$ (9.2 as compared to 9.6) in the extinction coefficient. For the β-lactoglobulin B solutions it leads to an error of $\sim 10\%$ in the molecular weight, $M_{w(c)}^a$, and of $\sim 100\%$ in the equilibrium constant, K_2.

where $R_a = M^a_{w(c)}/M_1$. When two virial coefficients are involved, the quantity $(1 - BM_1R_ac)$ becomes $(1 - BM_1R_ac - 2B_2M_1R_ac^2)$. Equation (III-4) is an algebraic modification of Equation (III-3).

(b) A least-squares procedure has been combined with the method of successive approximation, again with the presumption that a monomer–dimer reaction is involved. The actual data were analyzed by using a Wang desk-computer program based on a nonlinear least-squares fit. The measure of "goodness of fit" was defined as

$$\chi^2 \equiv \sum \left\{ \frac{1}{\sigma_i^2} [y_i - y(x_i)]^2 \right\} \tag{III-5}$$

This type of analysis has been described by Bevington (15). In it the σ_i are the uncertainties in the data points y_i. The optimum value of the parameters are obtained by minimization of χ^2 with respect to each of the parameters simultaneously. A grid-search technique was used in which the $M_1/M^a_{w(c)}$ values were calculated with a set of parameters and compared with the observed values by means of the χ^2 test. The parameters were then systematically changed, one at a time, to minimize χ^2 until the final minimum was reached.

(c) Chun et al. (10) have observed that when a single virial coefficient is involved the combination of Equation (III-1a) with the definition of $M^a_{n(c)}$ provides the interesting result

$$Z = \frac{2M_1}{M^a_{n(c)}} - \frac{M_1}{M^a_{w(c)}} = \frac{2M_1}{M_n} - \frac{M_1}{M_w} \tag{III-6}$$

Then, for the monomer–dimer reaction,

$$Z = f_1 + 1 - \frac{1}{2 - f_1} \tag{III-7}$$

in which f_1 is the actual weight fraction of monomer in the equilibrium mixture.

This equation may be solved by the quadratic formula to obtain values for f_1, since the quantity Z is available directly from the experimental data. Then, the plot of $(1 - f_1)/f_1$ versus cf_1 is linear and gives K_2 as slope.

The extension of this procedure to the monomer–dimer case where two virial coefficients are required to represent the data is achieved by the elimination of B_1M_1 and B_2M_1 from a combination of equations similar to equations (III-1a) and (III-1b), ones written with the two virial coefficients, in combination with a statement of Adams and Williams (16), as follows:

$$\ln f_1^a = \int_0^c \left(\frac{M_1}{M^a_{w(c)}} - 1 \right) d \ln c = \ln f_1 + B_1M_1c + B_2M_1c^2 \tag{III-8}$$

The quantity f_1^a is the apparent weight fraction of monomer at any concentration c in the cell at equilibrium.

The result is

$$\frac{6M_1}{M_{n(c)}^a} - \frac{M_1}{M_{w(c)}^a} - 2\ln f_1^a = 3 + 3f_1 - \frac{1}{2-f_1} - 2\ln f_1 \quad \text{(III-9)}$$

This equation in f_1 must be solved by successive approximations. Again, once f_1 has been determined, the equilibrium constant is readily calculable.

TREATMENT OF EXPERIMENTAL RESULTS

For the computations, starting with the $M_{w(c)}^a$ versus concentration data, it has been found that Methods a (Adams-Fujita-Van Holde-Rossetti) and b (Bevington) have been more successful in application than has Method c (Chun *et al.*)

TABLE III-1

VALUES OF CHARACTERISTIC CONSTANTS FOR DIMERI-
ZATION OF β-LACTOGLOBULIN B[a]

Temp. (°C)	K_2 (dl/g)	B_1M_1 (dl/g)	B_2M_1 (dl/g)
Method a [Equation (III-4)]			
5°	—	—	—
10°	54	0.105	—
15°	39	0.109	—
25°	23	0.110	—
35.5°	10.5	0.111	—
Method b [Equation (III-5)]			
5°	50	0.055	0.021
10°	44	0.085	0.007
15°	36	0.094	0.005

[a] pH 2.64, $I = 0.16$.

In Table III-1 are presented values of some of the characteristic constants, K_2, B_1M_1 (and B_2M_1) for the dimerization system, β-lactoglobulin B at pH 2.64, $I = 0.16$ and at several temperatures. The methods used for the computations are indicated.

It will be noted from this table that over the interval 10°–35.5°C the experimental data can be represented with the use of a single virial coefficient. At 10°C, both one and two virial coefficients could be adapted to reproduce the data within the precision of the experiment, but at 5°C it was not possible to find a combination of K_2 and $B_1 M_1$ to achieve satisfactory agreement. It is worthwhile to note that when two virial coefficients are used the values of K_2 are decreased in size.

In Table III-2 are presented observed and calculated apparent weight average molecular weight data at a selection of fringe numbers for the β-lactoglobulin B solution at $T = 35.5°C$. The calculated values agree quite well with the observed ones. The constants K_2 and BM_1 were obtained by using Equation (III-4).

TABLE III-2

COMPARISON OF OBSERVED AND COMPUTED VALUES OF $M_{w(c)}^a$ FOR THE β-LACTOGLOBULIN B SYSTEM AT 35.5°C[a]

Fringe number	$M_{w(c)}^a$ (obsd)	$M_{w(c)}^a$ (calcd)
7	26,435	26,500
9	27,030	26,996
11	27,410	27,313
13	27,720	27,511
15	27,870	27,629
17	27,920	27,686
19	27,880	27,700
21	27,820	27,677
23	27,740	27,628
25	27,650	27,557
27	27,520	27,469
29	27,370	27,367
47	25,970	26,127
63	24,870	24,871
71	24,270	24,249

[a] $K_2 = 10.5$, $B_1 M_1 = 0.111$. Data from Table III-1.

With good assurance of the essential purity of the protein and of the reaction type (monomer–dimer) the several basic thermodynamic functions, $\Delta G°$, $\Delta H°$, and $\Delta S°$ can be computed. They are evaluated from the single equilibrium constant for the formation of the intermolecular bond and its temperature variation by use of the conventional definitions.

The condition for chemical equilibrium is $\sum v_i \mu_i = 0$, where the v_i are stoichiometric mole numbers. From this statement it can be shown that

$$-\Delta G° = RT \ln K \qquad \text{(III-10)}$$

Here $\Delta G°$ is the standard Gibbs free energy change on the molar basis for the $2M_1 \rightleftarrows M_2$ reaction. The other symbols have their usual significance, with K, the equilibrium constant (written in terms of concentrations) as yet unspecified as to an association or dissociation reaction. Because the approximation of Adams and Fujita (1) for the activity coefficients has been used, $K = \prod_i c_i^{\nu_i}$.

The effect of temperature on K is described by the Gibbs-Helmholtz equation, to give ΔH°, the molal enthalpy change,

$$-\Delta H^\circ = \frac{R \, d \ln K}{d(1/T)} \tag{III-11}$$

The standard entropy change, ΔS°, is made available from the conventional definition

$$\Delta S^\circ = \frac{\Delta H^\circ - \Delta G^\circ}{T} \tag{III-12}$$

For the actual computations of these functions a conversion of concentration scales from grams per deciliter to moles per liter is required. We shall convert from our association constant, K_2, to values which describe the dissociation of dimer into monomer, a nonspontaneous reaction. The relation between the dissociation constant on the molar scale, K_D, and that on the g/dl scale, K_d, is

$$K_D = 20/M_1 K_d$$

Making use of the data the van't Hoff plot, $\ln K_D$ versus $1/T$ has been constructed. It is presented as Figure III-2. The data for the temperatures

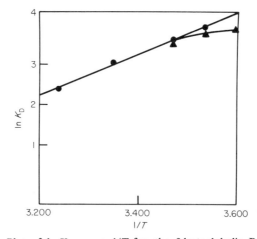

Figure III-2. Plot of $\ln K_D$ versus $1/T$ for the β-lactoglobulin B systems at several temperatures between 5° and 35.5°C. Units of $1/T$ have been multiplied by 10^3. For the computations the reaction has been written as a dissociation of dimer into monomer, a nonspontaneous reaction. ●, One virial coefficient; ▲, two virial coefficients. Reproduced from Visser et al. (6). Reprinted from *Biochemistry* 11, 2634 (1972). Copyright (1972) by the American Chemical Society. Reprinted by permission of the copyright owner.

10°, 15°, 25°, and 35.5° with one virial coefficient have been used for the computations. Data are for 25°.

$$\Delta G° = 5,840 \text{ cal/mole}$$
$$\Delta H° = 11,000 \text{ cal/mole}$$
$$\Delta S° = +17.5 \text{ e.u.}$$

The deviation from linearity in the $\ln K_D$ versus $1/T$ plot which appears by using dissociation constant data for the three parameter situations, temperatures 5° and 10°C, gives cause for concern because this van't Hoff plot is not a sensitive one. The digression from linearity is as yet unexplained. It might be tempting to suggest that a molecular conformation change in the protein is taking place at these low temperatures, but as of now this can be only conjecture, nothing more.

Lysozyme

For the protein system lysozyme at pH 7.0, $I = 0.20$, and $T = 25°C$ it will appear that in terms of the average deviation of the computed molecular weights from those actually measured, either the monomer–dimer or random self-association mechanism would be a reasonable choice. The two different mechanisms give nearly equivalent values for the dimerization constant. For the $M_{w(c)}^a/M_1$ versus c curve, it can be shown that for the monomer–dimer case the initial slope is $K_2 - BM_1$, while for the random reaction it is $2k - BM_1$ with k the intrinsic association constant. Since the two values obtained for BM_1 are small, $2k \cong K_2$.

Starting with Equation (III-3) Deonier has provided a method to determine the two parameters K_2 and BM_1 for a monomer–dimer reaction by an iterative procedure which is based upon a least-squares fit of the data to a linear equation. This new argument was presented (7, 7a) for application to the observed quantities in a sedimentation and chemical equilibrium experiment with lysozyme solutions under conditions for which the monomer–dimer model would appear to be adequate. For use with the experimental observations the Deonier procedure is to introduce a simplified notation, and so, under certain circumstances, to reduce the data to a linear plot from which the thermodynamic quantities are obtained. Again, the resulting parameters K_2 and BM_1 then may be inserted into Equation (III-3) to provide the $M_{w(c}^a$ versus c curve which is required by the use of the monomer–dimer model.

Actually, the possibility of the existence of reactions beyond the dimer stage must be investigated. For the simplest case, namely that in which a constant increment of free energy accompanies the addition of each monomer, an indefinite polymerization would be the result. We represent such a reaction by the following set of equations (*14, 17*). The species concentrations are here expressed on the molar basis, [c].

$$2M_1 \rightleftarrows M_2 \qquad K_a = [c_2]/[c_1]^2$$

$$M_1 + M_2 \rightleftarrows M_3 \qquad K_b = [c_3]/[c_2][c_1] \qquad \text{(III-13)}$$

$$M_i + M_1 \rightleftarrows M_{i+1} \qquad K_i = [c_{i+1}]/[c_i][c_1]$$

For constant free energy increment $K_a = K_b = K_c = \cdots K$; for the experiments and in defining the average molecular weights, the solutions must be made up on a weight per volume basis. Thus, if we wish to convert to an equilibrium constant on the grams per milliliter scale we would then have $k = 1000K/M_1$, with k being called the intrinsic equilibrium constant (again written for the association reaction).

The total concentration, c, is

$$c = c_1 + 2kc_1{}^2 + 3k^2c_1{}^3 + \cdots$$

$$= c_1/(1 - kc_1)^2 \qquad \text{when } kc_1 < 1 \qquad \text{(III-14)}$$

Then, for this same condition it can be shown that

$$\frac{M_1}{M^a_{w(c)}} = \frac{(1 - kc_1)}{(1 + kc_1)} + BM_1c \qquad \text{(III-15)}$$

and

$$\frac{M_1}{M^a_{n(c)}} = (1 - kc_1) + \frac{B}{2} M_1 c \qquad \text{(III-16)}$$

where $M^a_{n(c)}$, the number average molecular weight at concentration c, can be evaluated (numerical integration), again by using the relation

$$\frac{c_1 M_1}{M^a_{n(c)}} = \int_0^c \frac{M_1}{M^a_{w(c)'}} \, dc' \qquad \text{(III-17)}$$

Combination of Equations (III-15) and (III-16) gives

$$\frac{M_1}{M_{w(c)}} = \frac{\left(\dfrac{M_1}{M^a_{n(c)}} - \dfrac{B}{2} M_1 c \right)}{2 - \left(\dfrac{M_1}{M^a_{n(c)}} - \dfrac{B}{2} M_1 c \right)} + BM_1c \qquad \text{(III-18)}$$

Equation (III-18) can be solved for BM_1 by successive approximations, thus making the quantity $1 - kc_1$ subject to evaluation, by using Equation (III-16). The monomer concentration, computed by Equation (III-14), remains subject to the condition that $kc_1 < 1$.

The indefinite self-association may be also described by the equation (*14*)

$$\frac{R_a{}^2}{(1 - BM_1R_ac)^2} - 1 = 4kc \qquad \text{(III-19)}$$

This expression provides a very useful approach to analysis. We have noted that the intrinsic equilibrium constant, k, is expressed on a mass-based concentration scale. The method by which this equation is applied to experimental data has been described by Van Holde and associates (*14, 18*); it has been here used for the determination of the parameters $4k$ and BM_1 from the lysozyme experimental records instead of a method earlier elaborated by Adams (*9*).

EXPERIMENTAL RESULTS

Indefinite Self-Association

Sedimentation equilibrium experiments for the self-associating lysozyme in NaCl–cacodylate buffer at pH 7.0, ionic strength 0.20, and 25°C were conducted at 15 different initial protein concentrations. Values for the quantity $M_{w(c)}^a$, defined by Equation (III-2), plotted as a function of concentration c, were almost uniformly consistent. The experimental points, Figure III-3, are such as to suggest that the individual data points are generally within the

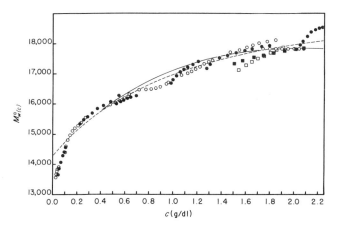

Figure III-3. Plot of $M_{w(c)}^a$ versus concentration of protein for the system lysozyme at pH 7.0 (cacodylate buffer), $I = 0.20$ and $T = 25°C$. Broken line: monomer–dimer association ($K_2 = 0.35$, $BM_1 = 0.02$); solid line: indefinite association ($2k = 0.40$, $BM_1 = 0.09$). Redrawn from Deonier and Williams (*7*). Reprinted with modification from *Biochemistry* **9**, 4260 (1970). Copyright (1970) by the American Chemical Society. Reprinted by permission of the copyright owner.

expected precision of the experiment, that the chemical equilibrium is rapidly reversible, and that the protein itself had been prepared in a sufficiently high degree of purity.

The values of $M^a_{w(c)}$ used for the numerical analyses were taken from an interpolation curve. The data for solutions of protein at concentrations less than 0.1 g/dl are suspect because to obtain them a 30 mm solution cell had to be employed. Therefore in this region, the main curve was interpolated to the ordinate at 14,300 g/mole, the monomer molecular weight as determined by amino acid analyses. Some values of $M^a_{w(c)}$ at several concentrations have been collected in Table III-3.

TABLE III-3

INTERPOLATED VALUES OF $M^a_{w(c)}$ AS A FUNCTION OF LYSOZYME CONCENTRATION

c (g/dl)	$M^a_{w(c)}$	c (g/dl)	$M^a_{w(c)}$
0.20	15,260	1.20	17,240
0.40	15,840	1.40	17,480
0.60	16,190	1.60	17,580
0.80	16,440	1.80	17,830
1.00	16,790	2.00	18,050
		2.20	18,240

As indicated, the Van Holde *et al.* (*14*) method of analysis has been applied to the data. Eleven points, taken from $M^a_{w(c)}$ values at regular concentration intervals from the interpolation curve, were fit to an appropriate linear equation by the use of the least-squares method to obtain the value, $BM_1 = 0.091$ dl/g. The average value of $4k$ which corresponds is 0.80 dl/g. The calculated values for $M^a_{w(c)}$ as a function of solute concentration which are required for the indefinite self-association, with the use of these two parameters, have been plotted as the solid line in Figure III-3.

Monomer–Dimer Equilibrium

The analysis of the $M^a_{w(c)}$ versus c data in terms of the monomer–dimer model was based on the Deonier (*7, 7a*) modifications of Equation (III-19) to which reference is made in the preceding section. The original article gives a description of the steps which lead to the evaluation of the quantities K_2 and BM_1, along with statements relative to the overall justification of the procedure. The numerical values found are $BM_1 = 0.02$ and $K_2 = 0.35$ dl/g.

When inserted into Equation (III-19), they generate the $M^a_{w(c)}$ versus c curve which is required for this reaction type; the resulting curve appears as the broken line in Figure III-3.

Originally it was believed that lysozyme in the specified solution was subject to the monomer–dimer type of self-association, even though there were indications in the literature that higher species might be involved. However, within the limits of the accuracy of the experimental data and up to protein concentrations of 2.2 g/dl, the random or indefinite self-association mechanism of reaction is about as well suited to the data as is the monomer–dimer stoichiometry. Calculated values of $M^a_{w(c)}/M_1$ as a function of concentration for the two reaction types closely approach each other, and the experimental data curve, as the concentration becomes smaller and smaller. The theoretical limiting slopes are $2k - BM_1$ for the random case and $K_2 - BM_1$ for the monomer–dimer situaton. The values of BM_1 are sufficiently small so that $2k \cong K_2$. At least then, we have a characteristic thermodynamic quantity. The actual data give $2k = 0.40$ dl/g and $K_2 = 0.35$ dl/g.

This interpretation can be justified in the following way. For the indefinite association the formation of dimer is described by the equation

$$c_2 = 2kc_1^2$$

with the quantity $2k$ performing the function of a dimerization constant. Thus the evaluation of the dimerization constant can be achieved to a reasonably good approximation even if the two alternative models represent quite different situations.

Chymotrypsinogen A

It has been found that chymotrypsinogen A undergoes a self-association under proper conditions. A quantitative evaluation of the sedimentation equilibrium data for our system (pH 7.9, $I = 0.03$, $T = 25°C$) indicates that the reaction type cannot be definitely assigned; although the indefinite self-association mechanism is perhaps more probable, the discrete mechanism of a monomer–dimer–trimer cannot be ruled out. The association constants, k in the one case, and K_2 and K_3 in the other, again provide a good estimate of one characteristic thermodynamic quantity.

The outline for the analysis of a system undergoing random association has already been written down. We present here in outline form an analysis for the monomer–dimer–trimer reaction, one in which the nonideality of solution behavior is again expressed in the terms BM_1. The concentrations are measured in grams per deciliter. [A more complete treatment has been presented by Adams (9).]

The nonideality term is evaluated by a process of successive approximation, using the equation

$$\frac{6cM_1}{M^a_{n(c)}} - 5c = 2cf_1^a \exp\left(-BM_1 c\right) + 3BM_1 c^2 - \frac{1}{\left(\dfrac{M_1}{cM^a_{w(c)}} - BM_1\right)} \tag{III-20}$$

The quantity $M^a_{n(c)}$ is available through Equation (III-17). The monomer concentration, c_1, is obtained from the modified Steiner equation and the formula

$$c_1 = cf_1^a \exp\left(-BM_1 c\right)$$

The equilibrium constants K_2 and K_3 have their sources in the two expressions,

$$K_2 c_1^2 = 3c - 2c_1 - \frac{1}{\left(\dfrac{M_1}{cM^a_{w(c)}} - BM_1\right)} \tag{III-21}$$

$$c = c_1 + K_2 c_1^2 + K_3 c_1^3 \tag{III-22}$$

The equilibrium constant K_3 characterizes the trimerization stage.

Thus, the curve $M_1/M^a_{w(c)}$ versus c may be constructed from the statement

$$\frac{M_1}{M^a_{w(c)}} = \frac{c}{c_1 + 2K_2 c_1^2 + 3K_3 c_1^3} + BM_1 c \tag{III-23}$$

for comparison with the experimental result.

EXPERIMENTAL RESULTS

In the performance of the sedimentation equilibrium experiments all known precautions were taken to ensure precision of result. The same assumptions, invariable values for the quantities $\bar{v}$ and dn/dc from species to species, and the same methods of analysis of the experimental records were continued in use. The basic statement for the sedimentation equilibrium, a definition of the quantity $M^a_{w(c)}$, Equation (III-2), is again applicable; its actual evaluation from the data is a subject of detailed description by Hancock and Williams (8).

The $M^a_{w(c)}$ versus c data for the chymotrypsinogen A are represented as points in Figure III-4, with the alternate open and filled circle regions being indicative of the several experiments which were performed at different initial protein concentrations. The reciprocal function plot, $M_1/M^a_{w(c)}$ versus c, is presented as Figure III-5, as an interpolated full line. The value for M_1 is independently known to be 25,000 from the amino acid composition and from the sedimentation equilibrium studies of LaBar (19).

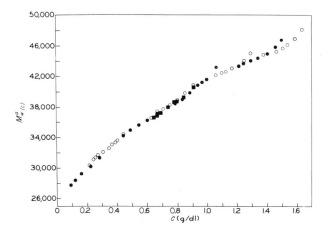

Figure III-4. . The behavior of the quantity $M^a_{w(c)}$ as a function of concentration for chymotrypsinogen A in solution at pH 7.9, $I = 0.03$, $T = 25°C$. Redrawn from Hancock and Williams (8). Reprinted with modification from *Biochemistry* **8**, 2598 (1969). Copyright (1969) by the American Chemical Society. Reprinted by permission of the copyright owner.

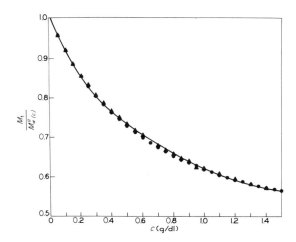

Figure III-5. Comparisons of experimental and calculated values of $M_1/M^a_{w(c)}$ as a function of concentration for the chymotrypsinogen A solution. Solid line: experiment. ▲, Calculated for indefinite association ($k = 0.50$ and $BM_1 = 0.04$); ●, calculated for monomer–dimer–trimer case ($K_2 = 0.91$, $K_3 = 0.84$, and $BM_1 = 0.01$). Redrawn from Hancock and Williams (8). Reprinted with modification from *Biochemistry* **8**, 2598 (1969). Copyright (1969) by the American Chemical Society. Reprinted by permission of the copyright owner.

The results of the calculations, using the random association model, are $2k = 0.99$ dl/g and $BM_1 = 0.04$ dl/g. These data were obtained by using Equations (III-18), (III-16) and (III-14). With them, the points which appear as filled triangles in Figure III-5 were obtained by application of Equation (III-15).

For the monomer–dimer–trimer model, it is necessary to use a negative value for the virial coefficient BM_1 ($= -0.01$) in Equation (III-23), an unrealistic situation. However, with this datum, $K_2 = 0.91$ and $K_3 = 0.84$ dl/g. It is seen from the values that $K_2 \cong (K_3)^{1/2}$ and that $2k \cong K_2$, as required.

The filled circles for $M_1/M_{w(c)}^a$ versus c in Figure III-5 have been computed by using Equation (III-23), with the three constants listed. The equivalence of all equilibrium constants was assumed in the derivation of the equations descriptive of the random association mechanism. From the way in which the equations for the monomer–dimer–trimer reactions were set up, and the near numerical equivalence of K_2 and $(K_3)^{1/2}$, one must expect that the closed [discrete (9)] association model will do almost as well in reproducing the experimental data as does the indefinite reaction. As with the lysozyme reaction, the definite choice of reaction mechanism is elusive.

However, the results again demonstrate that in the particular solution medium, a protein association reaction is taking place and that the process is readily reversible. The reaction association constants $2k$ (indefinite process) and K_2 (monomer–dimer–trimer reaction) are substantially equivalent, thus providing an important thermodynamic quantity.

Summary

An ever present problem in studies of the association of protein structures is the demonstration, first of all, that one really is dealing with a rapidly reversible self-association rather than with a heterogeneous nonassociating mixture. The average molecular weight data, plotted as a function of solute concentration, look superficially alike in the two situations. Several tests for the differentiation have been proposed. They often involve additional experiments of the transport variety, but observations of equilibrium may suffice for the purpose. The point has been discussed in a number of places in the literature. In at least one instance it has been recorded that data earlier understood to indicate self-association are now well accounted by a mixture of two inert proteins.

It has also been mentioned in several places in the text that the sedimentation equilibrium data, transformed to the $M_{w(c)}^a$ versus concentration curve, from a number of experiments at different initial solute concentrations must superimpose to form a single smooth curve in the self-association case.

Convincing demonstrations of the effect to the contrary of the presence of impurities are readily found. An inert impurity in the solution leads to complications when the attempt is made to establish the self-association mechanism, or even to decide whether such a reaction is actually involved.

Many proteins undergo reversible self-association in certain solution media. In some cases, the predominant reaction is one involving the formation of n-mer, in others an equilibrium distribution of polymers is the result, either in a limited way or in a random array. Also, there appear to be proteins which are capable of more than one of these types of self-association, depending upon the solution conditions; β-lactoglobulins A and B are representative examples. Still sought for their study are reliable and sufficiently sensitive physical methods to determine an average, or several average molecular weights. In this chapter we have given a record of progress, by using the sedimentation equilibrium method, in the characterization of the self-association reactions in solution of β-lactoglobulin B, lysozyme, and chymotrypsinogen A under well-defined conditions. Two items of information were sought: (a) determination of the self-association reaction type and (b) evaluation of the association constant (or constants) for the particular reaction.

In principle, the acquisition of accurate self-association equilibrium constant data is contingent upon the definitive, prior establishment of the reaction type. When the reaction tendency is small, it will be difficult to assign the mechanism, no matter what experiment is involved; there is nothing unique about a statement of this kind. In spite of this situation, it has been shown that some progress has been made in the numerical assignment of association constants, even though the methods for the complete identification of the stoichiometry remain imprecise. This fact is well demonstrated in the interpretation of the weight average molecular weight versus concentration data which have been provided by the ultracentrifuge for both the lysozyme and the chymotrypsinogen A systems.

For the self-association of lysozyme in solution, it has been claimed by others (a) that this protein is subject to a pH-dependent, reversible self-association into dimers, and (b) that the reaction goes beyond the dimer stage. In the two cases the actual solution conditions were not the same. In our own experience and with the protein in solution at one set of conditions (pH, ionic strength and temperature), it was seen that the data could be about as well accounted for by nonideal monomer–dimer behavior or by a random association process. Over the protein concentration range available (0.2 to 2.2 g/dl), the values of the quantity $M_{w(c)}^{a}/M_1$ computed for both models approach each other, and the experimental curve, as the protein concentration goes to zero. This situation can be readily appreciated by inspection of diagrams presented by Chun and Kim (20) in a recent article in which graphical analyses for several types of self-association appear. In spite of this situation, the evaluation of a reasonable equilibrium constant for the associative process was possible, since it was established that $2k \cong K_2$.

It might have been possible to eliminate one mechanism had we been able to extend the sedimentation equilibrium measurements over a wider concentration range without loss of precision. The overall precision of our low-speed, short-column experiments was 1%, with accuracy to within 2%. It is our opinion that the use of the high-speed or meniscus depletion procedure may not really improve matters, for while it might be more accurate at the lower concentrations, the reverse will be true at the higher concentrations.

The value of BM_1 ($= 0.09$ dl/g) obtained from the analysis by means of the equations descriptive of the random association is in quite good accord with values which have been measured for several other proteins of like molecular weight, in solution and at near zero charge and in high salt concentrations. This fact suggests that species higher than dimer may be present in relatively small amounts, with the sedimentation experiments failing to reveal their presence. Application of the monomer–dimer equations gives a BM_1 value which is appreciably smaller than would be expected (but still positive). We repeat, if there were some route by which an independent assignment of BM_1 could be made the overall situation could be greatly clarified and improved.

Although rather lengthy remarks have been made relative to it, there is nothing unique about the general lysozyme behavior. Similar statements could have been made about the behavior of chymotrypsinogen A and β-lactoglobulin B, each in solution under well-defined sets of conditions. The low-speed sedimentation equilibrium experiments were conducted to yield apparent weight average molecular weight versus protein concentration data at a single temperature for the chymotrypsinogen A solution and at five different temperatures in the interval 5° to 35°C for the β-lactoglobulin B acid solution. In these cases as well the concentration dependence of M_w^a could be ascribed to self-association, under nonideal conditions. The data were analyzed by using several, but by no means all, available theoretical formulations which are appropriate to well-defined reaction mechanisms.

It must be admitted that assignments of reaction type, followed by thermodynamic interpretation, leave much still to be desired. As one important item, improvements in both the precision and the accuracy of the basic sedimentation equilibrium information, a protein concentration versus distance curve over the cell, are clearly required.

Several methods are available for the determination of protein concentrations, but they must be applied with extreme care. As an example, if optical absorbance has been the method of choice, the extinction coefficient must be known with a high degree of accuracy, and immediately before the absorbance measurement it is important that the samples be filtered through millipore filters or otherwise controlled. One has the usual problems in the estimation of the ancillary constants, dn/dc and $(dp/dc)_\mu$. They, too, require accurate concentration assignment.

As a final comment, we reiterate that certain mass transport methods have been extensively applied in the detection and then the characterization of macromolecular interactions in terms of thermodynamic parameters. The theory of sedimentation and electrophoretic transport in self-associating systems has been elaborated especially by Gilbert and associates (*21*). Within its limitations, the Goldberg theory to describe the antigen–antibody reaction has been applied in the interpretation of proper sedimentation and electrophoretic transport diagrams to give the number of combining sites on an antibody molecule (*22, 23*), and to provide equilibrium constant data for the basic reaction $Ag + AgAb \rightleftarrows Ag_2Ab$ (*23*). This is an example of "mutual complementariness."

Whether an analysis of the equilibrium or of the transport type is selected, the path ahead seems certain to be long and arduous; nonetheless it is felt that the goal now can be more closely approached.

References

1. E. T. Adams, Jr., and H. Fujita, *in* "Ultracentrifugal Analysis in Theory and Experiment" (J. W. Williams, ed.), p. 119. Academic Press, New York, 1963.
2. A. Tiselius, *Z. Phys. Chem.* **124**, 449 (1926).
3. L. W. Nichol, J. L. Bethune, G. Kegeles and E. L. Hess, *in* "The Proteins" (H. Neurath, ed.), 2nd ed., Vol. 2, p. 305. Academic Press, New York, 1964.
4. E. F. Casassa and H. Eisenberg, *Advan. Protein Chem.* **19**, 287 (1964).
5. R. F. Steiner, *Arch. Biochem. Biophys.* **49**, 400 (1954).
6. J. Visser, R. C. Deonier, E. T. Adams, Jr., and J. W. Williams, *Biochemistry* **11**, 2634 (1972).
6a. D. A. Albright and J. W. Williams, *Biochemistry* **7**, 67 (1968).
7. R. C. Deonier and J. W. Williams, *Biochemistry* **9**, 4260 (1970).
7a. R. C. Deonier, Dissertation, University of Wisconsin (1970).
8. D. K. Hancock and J. W. Williams, *Biochemistry* **8**, 2598 (1969).
9. E. T. Adams, Jr., Fractions, No. 3, Spinco Division of Beckman Instruments, Palo Alto, California, 1967.
10. P. W. Chun, S. J. Kim, J. D. Williams, W. T. Cope, L-H. Tang and E. T. Adams, Jr., *Biopolymers* **11**, 197 (1972).
11. S. N. Timasheff and R. Townend, *J. Amer. Chem. Soc.* **83**, 470 (1961).
12. R. Townend, R. J. Winterbottom and S. N. Timasheff, *J. Amer. Chem. Soc.* **82**, 3161 (1960).
13. C. N. Pace, personal communication (1970).

14. K. E. Van Holde and G. P. Rossetti, *Biochemistry* **6**, 2189 (1967).
15. P. R. Bevington, "Data Reduction and Error Analysis for the Physical Sciences." McGraw-Hill, New York, 1969.
16. E. T. Adams, Jr., and J. W. Williams, *J. Amer. Chem. Soc.* **86**, 3454 (1964).
17. H-G. Elias and H. Lys, *Makromol. Chem.* **96**, 64 (1966); H-G. Elias and R. Bareiss, *Chima* **21**, 53 (1967).
18. K. E. Van Holde, G. P. Rossetti and R. D. Dyson, *Ann. N.Y. Acad. Sci.* **164**, 279 (1969).
19. F. E. LaBar, *Proc. Nat. Acad. Sci. U.S.* **54**, 31 (1965).
20. P. W. Chun and S. J. Kim, *Biochemistry* **9**, 1957 (1970).
21. G. A. Gilbert, *Proc. Roy. Soc., Ser. A* **250**, 377 (1959); **253**, 420 (1959).
22. R. J. Goldberg and J. W. Williams, *Discuss. Faraday Soc.* **13**, 224 (1953).
23. S. J. Singer and D. H. Campbell, *J. Amer. Chem. Soc.* **77**, 3499 and 3504 (1955).

CHAPTER **IV**

Sedimentation Analysis of a Multiple
Myeloma ɣG-Globulin

In Chapter III the combined sedimentation and chemical equilibrium behavior for three protein self-association reactions was considered. The point was made that unless means had been provided for the correction of non-ideality of solution behavior, it would have been impossible to evaluate the true equilibrium constants for the reactions discussed. With the use of the actual data, a concentration versus distance curve which was obtained from low-speed, short solution column, sedimentation equilibrium experiments, it became possible to take at least first steps in the assignment of mechanisms to the self-association reactions and to compute thermodynamic equilibrium constants for the dimerization steps.

Concurrently with the development of the theory and operational methods discussed in Chapter III, another but less comprehensive approach to the protein self-association problem, a scheme which makes use of a combination of sedimentation equilibrium and sedimentation velocity data, was devised. In its present form it is applicable only to an idealized dimerization; nevertheless the method gives a good approximation if the protein solutions are sufficiently dilute. It provides an apparent dimerization constant and limiting values for the sedimentation coefficients of monomer and dimer.

The approach seems to fit the sedimentation transport data over a much wider range of protein concentration than does the information from the equilibrium experiment. The reason would appear to be that the presence of polymers higher than dimer directly affects M_w, but if their amounts are still relatively small as compared with those of monomer and dimer (small values of an equilibrium constant) the position of the moving boundary gradient with time is relatively less influenced. This method has been applied in the study of the association behavior of a multiple myeloma ɣG-globulin and certain of its papain digestion products in two solvent media, cacodylate buffer and 8 M urea, at pH 7 and ionic strength 0.1, and at 25°C.

It is recognized that the analysis presented here has been used on relatively few subsequent occasions, and that other means to study protein self-association may be superior. However, it does constitute another approach and it has been used in the study of a γ-globulin, a subject of great interest.

The structure of the γG-immunoglobulin molecule itself has been investigated in detail. This molecule seems not to answer the usual description of "globular" molecule; further, the ellipsoid of revolution models cannot account for its known biological activities. It is composed of two identical halves, each of which contains two polypeptide chains, an H chain of molecular weight 50,000 and an L chain of molecular weight 25,000. The entity to which we shall refer as being the monomer unit contains four chains, two identical H and two identical L chains and has molecular weight of approximately 150,000. On cleavage of the molecule with hydrolytic enzymes and certain chemicals, it has been established that usually three main fragments are produced: two identical ones of the F_{ab} variety, each with an antibody site, and an F_c fragment. They all have substantially the same unit weight of about 50,000 daltons; they form a structure of three compact domains, with perhaps some flexibility of the units.

A number of models have been devised to account for the known behavior of the molecule. They can be classified as linear and Y forms. The attractive Edelman-Gally (1) model is an example of the former type. But there have been difficulties with it, and the Y-shaped structures of Noelken et al. (2) and of Valentine and Green (3) are presently in vogue. The electron micrographs of Valentine and Green do not of themselves show molecules of any one characteristic shape but they do indicate in a convincing way how with the Y model molecular associations to former dimers, trimers, etc., may take place. It is probably true that of all the proteins which have been studied for structure and composition, hemoglobin and the γG-immunoglobulins have received the most detailed attention.

During World War II our laboratory was engaged in research in aid of the large-scale production of the "γ-globulin antibodies" from human blood plasma. At this time it became of concern that their solutions could not be safely infused directly into the bloodstream; we became interested in whether molecular association reactions were a cause of the difficulty.

Although antibody activity has not been demonstrated for the myeloma γG-globulin, the molecules are structurally like those of the predominant immunoglobulins. The myeloma protein from a single individual is substantially homogeneous, thus making it a good material for the kind of study under consideration (4, 5).

Theory

The simplified theory for monomer–dimer self-association presented here makes use of a combination of the two kinds of sedimentation data.

SEDIMENTATION EQUILIBRIUM

For the rapidly reversible ideal equilibrium system, monomer–dimer,

$$2M_1/M^a_{w(c)} - 1 = 1/(1 + 4K_2c)^{1/2} \tag{IV-1}$$

For reference, the concentration $c = (c_a + c_b)/2$ is now adopted. It is the average concentration over the cell at sedimentation equilibrium, $\bar{c}$. The quantity $M^a_{w(c)}$ has its usual definition, $2RT/(1 - \bar{v}\rho)\omega^2 \, d \ln c/dr^2$. The dimerization constant is defined by the statement $K_2 = c_2/c_1^2$, with the total concentration being $c = c_1 + c_2$. Equation (IV-1) may be rearranged to give

$$[M^a_{w(\bar{c})}/(2M_1 - M^a_{w(\bar{c})})]^2 = 1 + 4K_2\bar{c} \tag{IV-2}$$

It requires that the quantity $[M^a_{w(\bar{c})}/(2M_1 - M^a_{w(\bar{c})})]^2$ should vary linearly with $\bar{c}$ and that the slope of the line should be equal to $4K_2$.

SEDIMENTATION VELOCITY

Fujita (6) has shown that the sedimentation coefficient for a monomer–dimer mixture may be expressed in the form

$$s = \frac{(s_1 + s_2K_2c_{av})}{(1 + K_2c_{av})} \tag{IV-3}$$

Here c_{av} is a somewhat different average concentration which is defined by the relation, $c_{av} = c_0(r_a/\bar{r}_H)^2$, in which c_0 is the initial protein concentration, r_a is the distance from the center of rotation to the meniscus, r_H is the radial position of the maximum in the refractive index gradient curve, and $\bar{r}_H$ is the arithmetic mean of the r_H values for the sedimentation boundary photographs taken at the beginning and at the end of the particular sedimentation experiment. The subscripts on s refer to the two protein species wich make up the single macromolecular component.

If it is assumed that both s_1 and s_2 have the same solute concentration dependence $(k_2 = k_1 = k_s)$, and if for clarity their limiting values are herein designated by superscript[0] for the purposes of this chapter,

$$s_1 = \frac{s_1{}^0}{(1 + k_sc_{av})} \tag{IV-4a}$$

$$s_2 = \frac{s_2{}^0}{(1 + k_sc_{av})} \tag{IV-4b}$$

then Equation (IV-3) becomes

$$s = \frac{(s_1{}^0 + s_2{}^0 K_2 c_{av})}{(1 + K_2 c_{av})(1 + k_s c_{av})} \tag{IV-5}$$

On rearrangement,

$$s = \frac{(s_1{}^0 - s_2{}^0)}{(1 + K_2 c_{av})(1 + k_s c_{av})} + \frac{s_2{}^0}{(1 + k_s c_{av})} \tag{IV-5a}$$

A plot for $1/s$ versus c_{av} should become linear as c_{av} becomes large, and the values of $s_2{}^0$ and k_s may be taken from the ordinate intercept and slope of this asymptotic line. Using another rearrangement,

$$s_2{}^0 - s(1 + k_s c_{av}) = \frac{(s_2{}^0 - s_1{}^0)}{(1 + K_2 c_{av})} \tag{IV-5b}$$

With values for $s_2{}^0$, k_s, and K_2 now available, the left-hand side of this equation can be evaluated, with the substitution of experimental values of s as a function of c_{av} and plotted as a function of $1/(1 + K_2 c_{av})$. This plot will be linear and pass through the origin, giving $(s_2{}^0 - s_1{}^0)$ as slope; $s_1{}^0$ is calculated from the latter quantity.

Experimental Results

The sedimentation equilibrium and transport behaviors which pertain to the solutions of the intact multiple myeloma γG-globulin and of two of its cleavage products are now considered. Suitable interpretations of the data for the intact protein molecule systems are more direct, and they will receive relatively greater attention. The protein itself, obtained from a single donor (HL), was a cryoglobulin that had been purified by three precipitations and differential centrifugations. The techniques used for the sedimentation analyses are substantially unchanged from those already considered in the preceding chapters and in Appendix A. For the equilibrium experiments, low rotor speeds and short solution columns were uniformly used. The velocity experiments were conducted in the conventional way, using double sector cells. A schlieren optical system usually provided the means to record the positions of the boundary gradient curves as a function of time, but on occasion the Rayleigh interferometer gave the boundary, c versus r, curve. The quantity $M_{w(\bar{c})}^a$ was taken at the midpoint between the meniscus and the bottom of the solution column from the plot of $\ln c$ versus r^2, corresponding very closely to the average protein concentration $\bar{c}$. Such data, $M_{w(\bar{c})}^a$ versus c, for the parent γG-globulin, are plotted in Figure IV-1.

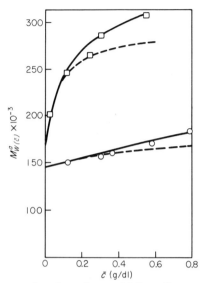

Figure IV-1. Concentration dependence of $M^a_{w(\bar{c})}$ of a parent myeloma γ-globulin at pH 7.0, $I = 0.1$, and $T = 25°C$ in two buffer solvent systems: $\bigcirc$——$\bigcirc$, cacodylate buffer. and $\square$——$\square$, 8 M urea. The dashed lines were computed by using the theory described in the text [Equation IV-1]. Redrawn from Kakiuchi and Williams (*4*).

It will be seen that in either of the solvents, 8 M urea or cacodylate buffer, the quantity $M^a_{w(\bar{c})}$ increases with increasing protein concentration. This increase in weight average molecular weight is pronounced in the urea solvent. The importance of accurate data in the region of very low solute concentration is quite apparent, and it is for this reason that Rayleigh interference optics were used to obtain the data at the two lowest concentrations. Even so, high accuracy cannot be claimed for these particular data points. However, all the information obtained is consistent with a monomer molecular weight of about 160,000. The corresponding value extrapolated from the data in the cacodylate buffer is about 145,000. Both figures are quite close to the datum 150,000, which is now generally accepted for the molecular weight of a γG-globulin.

The sedimentation coefficient information, to go with the apparent weight average molecular weight data, was obtained by standard procedures. These data have been assembled to construct Figure IV-2. The mobilities have been computed as $s_{25,w}$ values. The concentration is expressed as c_{av}, as defined above. The boundary gradient curves were quite symmetrical in the cacodylate buffer solutions, but they were somewhat skewed toward the region ahead of the maximum gradient in 8 M urea.

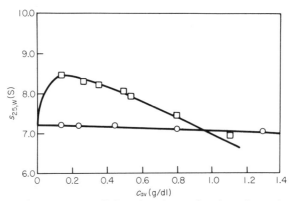

Figure IV-2. Sedimentation coefficient ($s_{25,w}$) as a function of protein concentration, c_{av}, for the parent γG-globulin in the two pH 7 buffers, cacodylate and 8 M urea. The solid lines represent computed values for comparison [Equation (IV-5)]. Redrawn from Kakiuchi and Williams (4).

The $s_{25,w}$ values in cacodylate buffer are substantially constant over the entire concentration range, with extrapolated value $s_{25,w}^0 = 7.2$ S, for zero solute concentration. The data in the urea system would appear to be fitted reasonably well by a straight line which would extrapolate to $s_{25,w} \cong 8.8$ S in the limit of zero concentration. However, to be consistent with the molecular weight data, an entirely different interpretation is required, and application of our theory will demonstrate that the sedimentation coefficient data for the protein in this solvent should follow a curve as shown by the solid line in Figure IV-2, one which requires a $s_{25,w}^0$ value which is identical with that found for the γG-globulin in the cacodylate buffer system.

Data which correspond have been obtained for two of the myeloma γG-globulin fragments—submolecules which have been separated from the papain –cysteine-reacted parent protein system by gel filtration. The two fragments have substantially the same molecular weight, about 50,000, but they differ greatly in their properties. One is the F_c portion, the unit with the two carbohydrate parts, and the other, a "univalent" antibody, which has been designated as F_{ab}. The solvents, cacodylate buffer and 8 M urea, are the same as described above; cf. Figure IV-1.

The apparent molecular weight data are presented in Figure IV-3.

The values observed for s_{25} in the urea solvent system have been converted to the water basis by using the viscosity ratio 1.70. For both fragments the coefficients are fitted by a single straight line, irrespective of solvent, with an intercept of $s_{25,w}^0 \cong 4.0$ S. For the urea case, k_s, defined by the equation $s_{25,w} = s_{25,w}^0 (1 - k_s c_{av})$, is 0.35 dl/g. For the cacodylate buffer, the limiting sedimentation coefficient is $s_{25,w}^0 = 4.1$ S, with k_s being not far from zero.

The transport data are also presented in graphical form Figure IV-4.

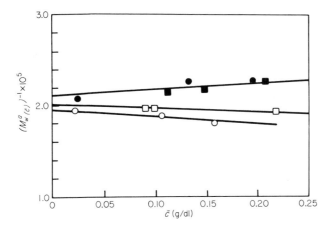

Figure IV-3. Inverse $M^a_{w(\bar{c})}$ as a function of the average protein concentration for the F_{ab} and F_c fragments of the multiple myeloma γ-globulin in the two pH 7 buffer–solvent systems: ●——●, fragment F_{ab} in the 8 M urea; ■——■, fragment F_c in 8 M urea; ○——○, fragment F_{ab} in cacodylate buffer; □——□, fragment F_c in cacodylate buffer. Redrawn from Iso and Williams (5).

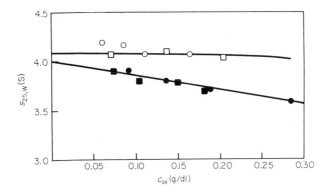

Figure IV-4. Concentration dependence of sedimentation coefficients for fragments F_{ab} and F_c in the two buffer–solvent systems. Upper curve (cacodylate buffer): ○——○, fragment F_{ab}; □——□, fragment F_c. Lower curve (8 M urea): ●——●, fragment F_{ab}; ■——■, fragment F_c. The upper solid line was constructed by using Equation (IV-7). Redrawn from Iso and Williams (5).

Correlation of the Data

In considering further the behavior of the parent γ-globulin, an attempt is made to show that the equilibrium data and the corresponding sedimentation velocity data are consistent with a rapid monomer–dimer self-association mechanism for dilute protein solutions, an effect which is pronounced in 8 M urea, and still readily observable in cacodylate buffer. To do so, it must be assumed that in these regions of low concentration, no polymers higher than dimers are formed, and that the effects of thermodynamic nonideality (other than that due to the protein self-association) may be neglected. It has to be recognized that this model is inadequate to cover the behavior over any extended range of concentration.

The equilibrium system is described by Equations (IV-1) and (IV-2). The latter form serves as a guide to the plots in Figures IV-5 and IV-6. From them

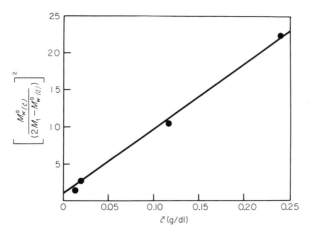

Figure IV-5.   The function $[(M^a_{w(\bar{c})}/(2M_1 - M_{w(\bar{c})})]^2$ plotted with $\bar{c}$ as abscissas for the γG-globulin when dissolved in 8 M urea solvent at pH 7. Redrawn from Kakiuchi and Williams (4).

it is found that K_2 is 21.5 dl/g for the reaction in the 8 M urea system and 0.33 dl/g in the aqueous cacodylate buffer. The plot of Figure IV-5 has not been extended beyond a protein concentration of 0.25 g/dl; at higher concentrations the same upward deviation is found which characterizes the reaction data for the system in cacodylate buffer, Figure IV-6. It is to be noted that the concentration scales in the two plots are greatly different. The presumption is that above a certain concentration higher polymers which coexist with monomer and dimer are formed in sufficient amount to cause the deviations from the straight line. These K_2 values were substituted into Equation (IV-1)

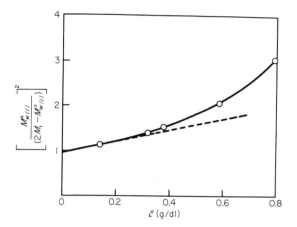

Figure IV-6. Plot of the quantity $[(M^a_{w(\bar c)}/(2M_1 - M^a_{w(\bar c)})]^2$ as a function of average concentration for the multiple myeloma γG-globulin in cacodylate buffer at pH 7. Redrawn from Kakiuchi and Williams (*4*).

in order to calculate $M^a_{w(\bar c)}$ as a function of c. These results, indicated by the dotted lines in Figure IV-1, agree quite well with experimental data in the concentration region below 0.25 g/dl; beyond this concentration range the presence of higher polymers is at once evident.

By using Equation (IV-5a), the plot of the $1/s_{25,w}$ versus c_{av} data for the 8 M urea, parent protein solution, Figure IV-7, is obtained. The curve approaches

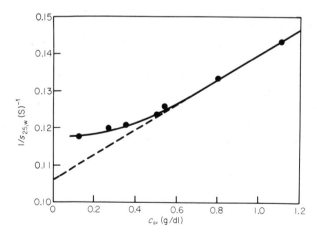

Figure IV-7. The function $1/s_{25,w}$ (in Svedberg units) plotted against the protein concentration c_{av} for the γG-globulin in 8 M urea solution at 25°C. Redrawn from Kakiuchi and Williams (*4*).

linearity as c_{av} becomes large, and values of $s_2{}^0$ and k_s are obtained from the intercept and slope of the asymptotic straight line. The dashed line gives for the dimer, $s_2{}^0 = 9.4$ S and $k_s = 0.32$ dl/g. The constants $s_2{}^0$, k_s and K_2 are by now available. Then, using Equation (IV-5b) as suggested, Figure IV-8

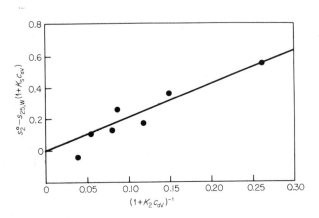

Figure IV-8. Plot of the function $s_2{}^0 - s_{25,w}(1 + k_s c_{av})$ versus $(1 + K_2 c_{av})^{-1}$ for the intact γG-globulin in 8 M urea solution. Redrawn from Kakiuchi and Williams (4).

was constructed. In this way, the monomer-limiting sedimentation coefficient is evaluated to be $s_1{}^0 = 7.2$ S. The points for the solid top line in Figure IV-2, descriptive of the transport behavior in the 8 M urea system were then calculated with the use of Equation (IV-5). Not only does the computed curve fit the experimental points well but it also converges at zero protein concentration to the expected $s_{25,w}^0$ value. This agreement suggests that the γG-globulin has essentially the same molecular conformation in both 8 M urea and cacodylate buffers. Even so, the velocity behavior in the urea solvent medium at very low concentrations should be confirmed by direct experiment, a very difficult measurement.

For the analysis of the velocity data in the cacodylate buffer, direct substitution has been made of the information $s_1{}^0 = 7.2$ S, $s_2{}^0 = 9.4$ S, and $K_2 = 0.33$ dl/g into Equation (IV-5) to find a value of k_s which will best fit the experimental data. The lower solid line of Figure IV-2 is the result, with $k_s = 0.09$ dl/g.

To begin the discussion of the sedimentation equilibrium behavior of the fragments F_{ab} and F_c in the two solvent systems, Equation (IV-1) is expanded to give

$$\frac{1}{M_{w(c)}^a} = \frac{1}{M_1} - \left(\frac{K_2}{M_1}\right)c + \cdots \tag{IV-6}$$

Thus, the initial slope of $1/M^{a}_{w(c)}$ with respect to c becomes negative when dimerization occurs. Furthermore, as was the case with the data, the plot of ln c versus r^2 was substantially linear and $M^{a}_{w(c)}$ may be replaced by M_{w}^{a}. Equation (IV-6) may be simplified to read

$$\frac{1}{M_{w}{}^{a}} = \frac{1}{M_1} - \left(\frac{K_2}{M_1}\right)\bar{c} + \cdots \tag{IV-6a}$$

At the same time we have replaced c by $\bar{c}$.

Application of Equation (IV-6a) to the sedimentation equilbrium data for the univalent antibody fragment, F_{ab}, and for the subunit, F_c, in the cacodylate buffer gives for the equilibrium constant, K_2, the values 0.35 and 0.20 dl/g, respectively. The monomer molecular weight in each case is approximately 50,000.

The sedimentation equilibrium data for these fragments when dissolved in 8 M urea buffer are not directly comparable with those characteristic of the parent protein molecule because p-chloromercuribenzoate was added to the solvent to repress any aggregative tendency. The curve obtained (Figure IV-3, upper line) is characteristic for a system in which the effects of solution nonideality are exerting an appreciable effect.

In Figure IV-4 are presented curves which describe the concentration dependence behavior of $s_{25,w}$, in S, for the two fragments, F_{ab} and F_c. In each instance the adjusted sedimentation coefficients for them are described by a single line, independent of solvent. The limiting sedimentation coefficients, $s^0_{25,w}$, are substantially the same in all four cases, combinations of two fragments and two solvents.

In the attempt to apply a modified Equation (IV-5) to the velocity data in cacodylic buffer for the computation of the constant k_s we may assign the parameters $s_1{}^0 = 4.1$ S, $s_2{}^0 = 6.5$ S, and $K_2 = 0.27$ dl/g. The $s_2{}^0$ value was estimated by using the simplest possible assumption, namely, that both monomer and dimer have spherical form in solution. The value of K_2 has been taken as the average of those which have been estimated from sedimentation equilibrium data, with solution nonideality effects being neglected to be sure.

Equation (IV-5) is written in the form

$$s = \frac{(s_1{}^0 + s_2{}^0 K_2 c)(1 - k_s c)}{(1 + K_2 c)} \tag{IV-7}$$

With the above parameters, a successive approximation procedure gives as best value, $k_s = 0.18$ dl/g. The calculated curve, shown as the upper line in Figure IV-4, seems to represent well the course of the experimental points, open circles and open squares.

Summary and Concluding Remarks

An examination of the literature reveals obvious disagreement about the molecular weights of the parent and subunit γ-globulin entities. It was the intention of this study to make more apparent the reasons for such discrepancies. In short, they are again largely due to failures to take into account concentration dependencies which derive from the protein self-association reactions.

Descriptions of some thermodynamic properties and of the sedimentation transport behavior in solution of an intact multiple myeloma γG-globulin and two of its enzymatic degradation products have been attempted. The solvents were cacodylate buffer and 8 M urea, both at pH 7, $I \cong 0.01$, and $T = 25°C$. Satisfactory interpretations of the data in both solvents for the parent macromolecule and for the fragments in cacodylate buffer require the use of a protein self-association mechanism. In all these cases it has been possible to estimate the limiting sedimentation coefficients for the monomer and dimer species, ones which are consistent with the observed sedimentation equilibrium data. In addition, the corresponding dimerization equilibrium constants have been evaluated. For the fragments in the urea system, the situation is not comparable since the associative reaction was suppressed by chemical addition.

A problem of related interest is whether the formation of proteins with known quaternary structures is different in kind from that of the rapidly reversible self-association reactions which take place in solution, such as those considered in the preceding chapter. The presumption is that we are dealing with a question of degree rather than of kind. The measure of the degree is the same—the association reaction equilibrium constant. This quantity will be very large to give the quaternary structure, with the reaction having moved well along toward completion. The magnitude of this constant may be thought to depend upon the conformation of the subunits as they have been modified by evolution.

The determination of the number of subunits per protein molecule is also pertinent. It will not always be possible to crystallize certain proteins in which the biological interest is great, and thermodynamic solution methods may be indicated for application. It was the accuracy of the molecular weight data made available by the ultracentrifuge which early made it possible to establish the existence of subunits in the multichain proteins and to determine their number per parent molecule. Unfortunately, in order to dissociate the protein into the subunits, it is often necessary to resort to the use of complex solvent mixtures. Thus, the interpretation of the data may involve more intricate mathematical processes, ones which have been sometimes not fully appreciated.

This brings us to a concluding thought. It has been stated, in effect, that mathematics is the only science by which anything can be proven. True or not, this discipline does provide the means by which experiments are designed and by which the data therefrom are evaluated and interpreted. This language guides the transformation from one type of distribution to another. It has greatly expanded the usefulness of the ultracentrifuge by providing the change-over from analyses largely restricted for application to the two-component, ideal solutions, to the treatment of multicomponent systems, the constituents of which display thermodynamic nonidealities. For the ultracentrifuge, theory and practice go together; there must be constant interaction between them for growth.

References

1. G. M. Edelman and J. A. Gally, *Proc. Nat. Acad. Sci. U.S.* **51**, 846 (1964); however, cf. G. M. Edelman and W. E. Gall, *Ann. Rev. Biochem.* **38**, 415 (1969).
2. M. E. Noelken, C. A. Nelson, C. E. Buckley, III, and C. Tanford, *J. Biol. Chem.* **240**, 218 (1965).
3. R. C. Valentine and N. M. Green, *J. Mol. Biol.* **27**, 615 (1967); R. C. Valentine, *Gamma Globulins, Proc. 1967 Nobel Symp.*, 3rd, p. 251 (1967).
4. K. Kakiuchi and J. W. Williams, *J. Biol. Chem.* **241**, 2781 (1966).
5. N. Iso and J. W. Williams, *J. Biol. Chem.* **241**, 2787 (1966).
6. H. Fujita, "The Mathematical Theory of Sedimentation Analysis," p. 205. Academic Press, New York, 1962.

PART 3

APPENDIXES

A Brief Introduction to the Theory of Ultracentrifugal Analysis

We offer here a brief introduction to the background and general theory of ultracentrifugal analysis. The account is intended to aid the reader in the study of the topics under consideration, but at the same time it is hoped that reference to more comprehensive treatises will not be thereby discouraged.

The solutes involved are macromolecules, either proteins or organic high polymers. For each type of system there are restrictive factors.

(1) Proteins, while homogeneous, are amphoteric polyelectrolytes; thus charge effects may complicate the equations by which molecular weights, sedimentation coefficients, etc., are evaluated. However, these charge effects may be repressed by the presence of added simple supporting electrolyte and by working with solutions of pH in the isoelectric region. Thus, the excluded volume may become the main cause of solution nonideality and it is usually quite a small effect. Under certain circumstances, many proteins undergo self-association in solution and for the characterization of the monomers it becomes necessary to find conditions of pH, ionic strength, temperature, and solute concentration where such reactions are largely absent. Where present, the mechanism of the association reaction may be investigated and in certain cases, assigned.

(2) In contrast with the proteins, the organic high polymers are generally polydisperse but electrically neutral. They may require use of organic solvents, sometimes compressible under the conditions required for the experiment. The problems which go with the interpretation of the experimental data for these systems, often further complicated by pronounced intermolecular interactions, thus may become formidable. In recent years, advances in theory have served to alleviate some of the difficulties, indeed to the extent that the

sedimentation methods are now becoming quite successful in the detailed description of certain types of polymer systems, at least.

Of primary concern in ultracentrifugal analysis is the geometry of the solution cell. For the transport experiment, it is required that the path of solute sedimentation be along radial lines in order to avoid convective disturbances in the solution during the process. Cells of other shapes (rectangular, for example) may be used for observations of the equilibrium condition, but it is practical even here to use the same sector-shaped cells, with theory to conform.

The equations that will be set down as descriptive of the two sedimentation behaviors are written in cylindrical coordinates; a sector angle, θ; a cell depth, h (a distance which is parallel to the axis of rotation); and a radial distance, measured by r (Figure A-1). The solute concentration in the cell at a given time is a function only of radial distance; thus the other coordinates do not ordinarily appear in the final equations.

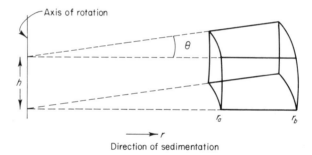

Figure A-1. Diagrammatic sketch to show the location and geometry of the solution cell in the ultracentrifuge. The cell itself is of sector shape.

r = Radial distance from the center of rotation, with r_a and r_b the meniscus and cell bottom positions (air interface and layering fluid not shown).

h = Cell height, parallel to the axis of revolution.

θ = Sector angle, of magnitude to produce sedimentation along radial lines.

Basic Principles

Making use of a statement of the conservation of mass, the amount, g_i, of a solute in the cell is expressed as

$$g_i = \int_a^b c_i (dV/dr)\, dr = \text{constant} \tag{A-1}$$

where c_i is the solute concentration on a gram per volume scale, at radial distance r; dV/dr is the rate of change of cell volume, V, with r; and the top and bottom of the cell are denoted by the distances a and b.

The cell area, normal to a radius, $A = kr$, is given by dV/dr, and we have

$$g_i = \int_a^b c_i \, dV = k \int_a^b rc_i \, dr \tag{A-2}$$

with

$$\int_a^b rc_i \, dr = (c_i)_0 \int_a^b r \, dr = (c_i)_0 (b^2 - a^2)/2 \tag{A-3}$$

The constant k depends upon the magnitude of the sector angle and the cell thickness. The initial solute concentration of the ith component is denoted by $(c_i)_0$.

THE CONTINUITY EQUATION

Another useful statement of the conservation of mass gives (in the absence of chemical reactions)

$$\frac{dg_i}{dt} = (krJ_i)_{r_1} - (krJ_i)_{r_2} = \frac{\partial}{\partial t} \int_{r_1}^{r_2} krc_i \, dr \tag{A-4}$$

Stated in words, the mass g_i of a solute i, flowing in across a boundary at radial distance r_1 minus the mass flowing out at another radial distance, r_2, is equal to the time rate of change of its mass remaining in the volume element bounded by r_1 and r_2. The symbol J_i represents the solute flux and is given by the product $c_i v_i$, where v_i is the velocity in the field of the ith kind of molecule.

In differential form, this equation is written

$$\left(\frac{\partial c_i}{\partial t} \right)_r = -\frac{1}{r} \left[\frac{\partial (rJ_i)}{\partial r} \right]_t \tag{A-5}$$

to give what is called the "continuity equation". With adequate flow equations, it leads at once to the differential equation of the ultracentrifuge for the two-component system, and to a set of coupled equations for multi-component systems. These flow equations are now derived by the methods of thermodynamics of irreversible processes (1,2). The common procedures for the determination of a sedimentation coefficient make use of solutions of this differential equation for the two-component system.

Sedimentation Equilibrium

In principle, the essential working equations for the interpretation of sedimentation equilibrium experiments could have been obtained by setting, in Equation (A-5), $(\partial c_i/\partial t)_r = 0$, for the equilibrium condition. This is equivalent to the assumption that diffusional and sedimentation flows, taken for the present to be additive in the ultracentrifuge process, are equal and opposite at every point in the cell. It is a more direct approach to utilize the methods of classical thermodynamics, following Gibbs. Accordingly, at equilibrium, the total potential of each component is independent of position in the ultracentrifugal field. This is the route that we shall follow.

Consider a system of several phases in thermal equilibrium, for which there exists the possibility of free exchange of chemical substance between neighboring phases. Applicable to this system is the Gibbs equation

$$dU = T\,dS - P\,dV + \sum_i \mu_i\,dn_i$$

in which n_i is the number of moles of component i in the system and μ_i, the chemical potential of i, is the energy brought into the system at constant volume, entropy, and at constant composition of all other components per mole of added i (the other symbols have their usual significance):

$$\mu_i = \left(\frac{\partial U}{\partial n_i}\right)_{S,V,n_j} = \left[\left(\frac{\partial G}{\partial n_i}\right)_{T,P,n_j}\right] \tag{A-6}$$

where μ_i is the chemical potential of the ith component on a molar basis.

In the presence of applied external fields, it is the total potential $\bar{\mu}_i$, which is constant in all phases at equilibrium, thus,

$$\bar{\mu}_i = \mu_i - M_i\omega^2 r^2/2 = \text{constant in an ultracentrifugal field} \tag{A-7}$$

$$\bar{\mu}_i = \mu_i - M_i\omega^2 r^2/2 + z_i\varepsilon\psi = \text{constant in a combined ultracentrifugal and electrical field, etc.}$$

In these definitions, M_i is the molecular weight of component i, ω is the angular speed of the rotor, ψ is the electrical potential of the phase, ε is the charge of a mole of electrons, and z_i is the valence of i per mole, in chemical units. When the system under consideration contains only neutral molecules, the electrochemical potential term does not appear.

The chemical potential of a nonionizing solute is related to a volume-based concentration, c_i, in grams per cubic centimeter, as follows:

$$\mu_i = (\mu_i)_c^0 + RT\ln y_i c_i$$

In this definition $(\mu_i)_c^0$ is the reference chemical potential of solute species i, with concentration expressed on the c scale, and y_i is an activity coefficient again with respect to the c concentration scale. In adopting the c scale, the equations as written throughout this section are restricted to incompressible systems (2). On the molality scale (and activity coefficient to conform) this restriction is removed. The equations for $\bar{\mu}_i$ and μ_i, though written here on a per mole of component i basis, could have been expressed on the per gram basis and indeed there may be advantages in doing so, especially when the equations for sedimentation transport are generalized.

BINARY SYSTEMS (Incompressible Solutions of Neutral Molecules)

The criteria for equilibrium are

$$T = \text{constant} \tag{A-8}$$

$$\frac{dP}{dr} - \rho\omega^2 r = 0 \tag{A-9}$$

$$\frac{d\bar{\mu}_2}{dr} = \frac{d\mu_2}{dr} - M_2\omega^2 r = 0 \tag{A-10}$$

The quantities T, P, and ρ are the temperature, pressure, and solution density, respectively, at the radial distance r. The molecular weights of the components 1 and 2 are M_1 and M_2, with the subscript 2 being used not only here, but throughout, for the macromolecular component. The predominant solvent is specified as component 1.

At constant temperature the chemical potential of any component is a function of the pressure and the concentration of the component. Then,

$$\frac{d\mu_2}{dr} = \left(\frac{\partial\mu_2}{\partial P}\right)_{c_2}\frac{dP}{dr} + \left(\frac{\partial\mu_2}{\partial c_2}\right)_P\frac{dc_2}{dr} \tag{A-11}$$

Also,

$$\left(\frac{\partial\mu_2}{\partial P}\right)_{c_2} = \bar{V}_2 = M_2\bar{v}_2 \tag{A-12}$$

where $\bar{V}_2$ and $\bar{v}_2$ are the partial molal volume and partial specific volume of component 2, respectively.

Combination of the Equations (A-9) through (A-12) gives

$$M_2(1 - \bar{v}_2\rho)\omega^2 r = \left(\frac{\partial\mu_2}{\partial c_2}\right)_P\frac{dc_2}{dr} \tag{A-13}$$

With the definition of the chemical potential, Equation (A-7), we now write

$$M_2(1 - \bar{v}_2\rho)\omega^2 r = \frac{RT}{c_2}\frac{dc_2}{dr} \quad \text{(ideal solution)} \tag{A-14a}$$

$$M_2(1 - \bar{v}_2\rho)\omega^2 r = \frac{RT}{c_2}\left[1 + c_2\left(\frac{\partial \ln y_2}{\partial c_2}\right)_{T,P}\right]\frac{dc_2}{dr} \quad \text{(nonideal solution)} \tag{A-14b}$$

To integrate Equation (A-14 b) for c_2, certain assumptions are required, such as

(a) $\ln y_2 = B_2 M_2 c_2 + \cdots$

(b) $\rho = \rho_1 + (1 - \bar{v}_2 \rho_1)c_4$

In (a) the pressure dependence of y_2, the solute activity coefficient, is neglected; for ρ and $\bar{v}_2$ any pressure dependence is assumed to be absent; an incompressible solution. Then,

$$\frac{M_2(1 - \bar{v}_2 \rho_1)\omega^2 r c_2}{RT} = [1 + (B_2 M_2 + \bar{v}_2)c_2 + \cdots] \frac{dc_2}{dr} \quad (A-15)$$

Then, by using Equation (A-3) the necessary boundary condition for the use of Equation (A-15) is obtained. By its integration we now have

$$\frac{1}{M_2{}^a} = \frac{1}{M_2} + B_2{}' \frac{[c_2(r_b) + c_2(r_a)]}{2} + \cdots \quad (A-16)$$

where

$$M_2{}^a = \frac{2RT}{(1 - \bar{v}_2 \rho_1)(r_b{}^2 - r_a{}^2)\omega^2} \left[\frac{c_2(r_b) - c_2(r_a)}{(c_2)_0} \right] = \begin{array}{l} \text{apparent solute} \\ \text{molecular weight} \end{array}$$

$$B_2{}' = B_2 + (\bar{v}_2/M_2)$$

Equation (A-16) takes a very simple form when it is noted that $\frac{1}{2}[c_2(r_b) + c_2(r_a)]$ is the average concentration over the cell; it will be designated as $\bar{c}$.

The result, Equation (A-16), indicates that a plot of $1/M_2{}^a$ versus the concentration parameter $\bar{c}$ should be a straight line in the lower concentration regions (since the higher terms in the expansion for $\ln y_2$ have been neglected, a condition which is often compatible with experiment). The intercept as $c_2 \to 0$ and the slope of the line provide values of $1/M_2$ and $B_2{}'$, respectively. The concentration difference is a directly measurable quantity when the Rayleigh integral fringe method is used.

The analogs of Equation (A-16) for osmotic pressure and light scattering of an incompressible binary solution at a fixed temperature are

$$\frac{\pi}{RT(c_2)_0} = \frac{1}{M_2} + \frac{1}{2}\left(B_2 + \frac{\bar{v}_2}{M_2}\right)(c_2)_0 + \cdots$$

$$\frac{H(c_2)_0}{\tau} = \frac{1}{M_2} + \left(B_2 + \frac{\bar{v}_2}{M_2}\right)(c_2)_0 + \cdots$$

The concentration, $(c_2)_0$, is that of the particular solution, π is its osmotic pressure, and τ is the excess turbidity due to the presence of the solute, component 2. The quantity H is familiar as the light-scattering factor. The osmotic pressure equation follows from the application of the Gibbs-Duhem relation at constant temperature and pressure. The derivation of the equation for light scattering involves fluctuation theory.

For the "low-speed" sedimentation equilibrium experiment, we set down in explicit form equalities which have been already used in the interpretation of Equation (A-16)

$$\frac{c_2(r_b) - c_2(r_a)}{(c_2)_0} = \frac{\Delta c_2}{(c_2)_0} = \frac{M_2(1 - \bar{v}_2\rho_1)\omega^2(b^2 - a^2)}{2RT} = \lambda M_2 \quad (A-17)$$

For simplicity, $r_b^2 - r_a^2$ has been replaced by $b^2 - a^2$. In this way, it is necessary to evaluate the concentration at the meniscus, and the difference in concentration between meniscus, a, and bottom of the solution column, b. The original concentration is also required. In the proper experiment, the ratio $\Delta c_2/(c_2)_0$ is made available with quite good accuracy.

The equivalent of the expressions (A-14) for the distribution of the macromolecular component at sedimentation equilibrium appears in many forms in the literature. Thus, for example, one finds the convenient and useful shorthand statements:

$$\frac{d \ln c_2}{d(r^2)} = AM_2 \qquad \text{(ideal solution)} \qquad (A-14c)$$

$$\frac{d \ln a_2}{d(r^2)} = AM_2 \qquad \text{(nonideal solution)} \qquad (A-14d)$$

In them,

$$A = \frac{(1 - \bar{v}_2\rho_1)\omega^2}{2RT}$$

TERNARY SYSTEMS (Incompressible Solutions of Low Molecular Weight Components 1, 3, and the Macromolecular Component 2)

In certain experimental situations, it is necessary to add a second component to the main "solvent" component, one which may be itself subject to appreciable redistribution at sedimentation equilibrium. This is a situation which must be taken into consideration in the interpretation of the data. An extrapolation of the apparent molecular weight of the macromolecular component will not generally lead to the true molecular weight of the unsolvated macromolecule if the equations applicable to two-component systems, already written down, are employed.

The chemical potential of any component is now regarded to be a function of the pressure and of the concentration of any two of the components, and two (of three similar) equations are required to describe the equilibrium condition. Noting again that for exactness the molality concentration scale would have been preferable, they are

$$M_2(1 - \bar{v}_2\rho)\omega^2 r = \left(\frac{\partial\mu_2}{\partial c_2}\right)_{c_3,P}\frac{dc_2}{dr} + \left(\frac{\partial\mu_2}{\partial c_3}\right)_{c_2,P}\frac{dc_3}{dr} \qquad \text{(A-18)}$$

$$M_3(1 - \bar{v}_3\rho)\omega^2 r = \left(\frac{\partial\mu_3}{\partial c_2}\right)_{c_3,P}\frac{dc_2}{dr} + \left(\frac{\partial\mu_3}{\partial c_3}\right)_{c_2,P}\frac{dc_3}{dr} \qquad \text{(A-19)}$$

They follow from statements of the criteria for equilibrium, comparable to Equations (A-8), (A-9), and (A-10), except that now two (i.e., $q - 1$) equations which correspond to Equation (A-10) are required (q number of components).

$$\frac{d\mu_2}{dr} - M_2\omega^2 r = 0$$

$$\frac{d\mu_3}{dr} - M_3\omega^2 r = 0$$

If Equations (A-18) and (A-19) are solved simultaneously, a rather cumbersome expression is obtained for the apparent molecular weight, M_2^a. We shall require only the limiting expression for the case of vanishingly small macromolecular concentration, which is

$$\lim_{c_2 \to 0} M_2^a = M_2\left\{1 + \Gamma\frac{M_3(1 - \bar{v}_3\rho)}{M_2(1 - \bar{v}_2\rho)}\right\} \qquad \text{(A-20)}$$

$$\Gamma = -\frac{(\partial\mu_3/\partial m_2)_{m_3,P}}{(\partial\mu_3/\partial m_3)_{m_2,P}} = \left(\frac{\partial m_3}{\partial m_2}\right)_{\mu_3,P} \cong \left(\frac{\partial c_3}{\partial c_2}\right)_{\mu_3,P} \qquad \text{(A-21)}$$

The protein physical chemist is concerned with the ternary system H_2O, 1; macroelectrolyte PX_z, 2, and simple electrolyte BX, 3. If, in a binary system, the solute dissociates to produce $z + 1$ ions, the apparent molecular weight which is obtained is $1/(z + 1)$ times the true value. However, the complete removal of extraneous electrolyte is exceedingly difficult and it has become general practice to add an excess of low molecular weight salt (supporting electrolyte) to the system in order to make very small the electrostatic charge effects. It has already been indicated that the redistribution of a solvent component must be taken into account to obtain a true macromolecule weight, even though the apparent molecular weight has been extrapolated to infinite dilution.

In the performance of a molecular weight determination of a protein, the multicomponent system is dialyzed to equilibrium against the salt solution in which the macromolecule has been dissolved. If the definition of components

2 and 3 as strong electrolytes is retained, it can be shown that by using the conditions for Donnan membrane equilibrium,

$$\lim_{c_2 \to 0} \Gamma = -\tfrac{1}{2}z$$

and

$$\lim_{c_2 \to 0} M_2^{\text{a}} = M_2\left[1 - \tfrac{1}{2}z \frac{M_3(1 - \bar{v}_3\rho)}{M_2(1 - \bar{v}_2\rho)} \right] \tag{A-22}$$

Thus, a residual charge effect persists even at zero protein concentration and the molecular weight obtained on extrapolation will be somewhat less than the true value.

This same result was obtained in more direct fashion by Lamm in 1944 (3); the failure to abide by this early and very important teaching has been the cause of much later confusion in the literature. To conform to a greater extent to the notation used by Lamm, Equation (A-22) may be written as

$$\lim_{c_2 \to 0} \frac{d \ln c_2}{d(r^2)} = A_{\text{PX}_z} M_{\text{PX}_z} - \tfrac{1}{2} z A_{\text{BX}} M_{\text{BX}} \tag{A-23}$$

In a theoretical study of osmotic pressure, this result of Lamm may have suggested to Scatchard (4) that the macromolecular component could be redefined as $\text{PX}_z - \tfrac{1}{2}z\,\text{BX}$, a neutral component. Following an auspicious start by Vrij in his Utrecht dissertation (1959), Eisenberg (5) and Casassa and Eisenberg (6) have demonstrated in detail how the equations for sedimentation equilibrium in typical monodisperse protein systems, in the presence of supporting electrolyte, when formulated in terms of certain refractive index and density derivatives, reduce in form to the simple equations for two-component systems. Through their use, they provide directly quite accurate protein molecular weight data. Their analyses show that it is not necessary to resort to a redefinition of components to achieve the favorable result. Actually, the new equations are applicable to any ternary system, whether or not the solute is ionizable, provided that these density derivatives can be made amenable to experimental determination.

MULTICOMPONENT SYSTEMS

Here again classical thermodynamics provides the description of the sedimentation equilibrium. As before, we regard the system to be made up of a continuous sequence of phases of fixed volume and in the direction of the ultracentrifugal field. The criteria for equilibrium for each component are unchanged as compared to those already used in connection with the analysis of two- and three-component systems. We assume that the density of the solution and the partial specific volumes of the several components are independent of pressure and composition. In those cases which will be of interest to us, the organic high polymers, the macromolecular components are made up of units which differ only in molecular chain length, and it can be reasonably assumed that the partial specific volumes are the same for all such components.

The chemical potential of each macromolecular component will be a function of the pressure and of the concentration of all components. Thus, instead of the single Equation (A-13), a set of $(q - 1)$ equations is required to describe the equilibrium. The set is given in condensed form, as follows:

$$M_i(1 - \bar{v}_i\rho)\omega^2 r = \sum_{k=2}^{q}\left(\frac{\partial \mu_i}{\partial c_k(r)}\right)_{T,P,c_j}\frac{dc_k(r)}{dr} \qquad (i = 2, \ldots, q) \quad \text{(A-24)}$$

Making use of the definition of the chemical potential of component i we have

$$\frac{M_i(1 - \bar{v}_i\rho)\omega^2 r}{RT} = \frac{1}{c_i(r)}\frac{dc_i(r)}{dr} + \sum_{k=2}^{n}\left(\frac{\partial \ln y_i}{\partial c_k(r)}\right)_{T,P,c_j}\frac{dc_k(r)}{dr}$$

The symbol $c_i(r)$ refers to the concentration of the ith component at cell position r at equilibrium. If the solutions are sufficiently dilute, the solution density, ρ, may be replaced by the density of the solvent, ρ_1. Then, with constant $\bar{v}$ for all solute components

$$\frac{M_i(1 - \bar{v}\rho_1)\omega^2 rc_i}{RT} = \frac{dc_i}{dr} + c_i\sum_{k=2}^{q}\left(\frac{\partial \ln y_i}{\partial c_k}\right)_{T,P,c_j}\frac{dc_k}{dr} \qquad \text{(A-25)}$$

The logarithm of y_i is expressed in powers of c_k, in the form (dilute solutions)

$$\ln y_i = M_i\sum_{j=2}^{q} B_{ij}c_j + \cdots \qquad \text{(A-26)}$$

and

$$\frac{M_i(1 - \bar{v}\rho_1)\omega^2 rc_i}{RT} = \frac{dc_i}{dr} + M_ic_i\sum_{k=2}^{q} B_{ik}\frac{dc_k}{dr} \qquad \text{(A-27)}$$

The B_{ik} are very complicated functions of M_i and M_k, and much effort has been expended to simplify the situation, so that simpler equations can be deduced for use with experimental data. And there still remains the problem of which concentration-dependent variable should be used in connection with the experimental data.

In an early attempt to answer the question, Wales, Adler, and Van Holde (7) had proposed the following empirical expression for use with multi-component systems.

$$\frac{1}{M_w{}^a} = \frac{1}{M_w} + B(c_2)_0$$

It was to be used in situations which answered the description, "a small cell, or its equivalent, a not too great separation of components." In present day parlance, this restriction would have read "low speeds and short solution columns." An analog of Equation (A-16) in $\bar{c}$ is to be preferred; it teaches that for a multicomponent nonideal system the plot to be used should be one of $(M_w^a)^{-1}$ versus $\bar{c}$ for higher accuracy.

In Chapter I we discussed at some length the question of the conditions and extent to which Equation (A-16) may be applied in the evaluation of data for nonideal solutions of polydispersed solutes, merely by the replacement of M_2 with the weight average molecular weight of the solute and by regarding B as a parameter which may be compared to the light-scattering second virial coefficient.

It was seen that with the use of certain approximations and appropriate experimental conditions, the behavior of a polydisperse nonideal solution may be treated as if a nonideal two-component system were involved, with the molecular weights being values which have been averaged over the cell in well-defined ways.

For the polymeric systems, it is the weight average molecular weight which is most readily available. A simplified argument follows:

$$\frac{1}{r} \sum_{i=2}^{q} \frac{dc_i}{dr} = \frac{\omega^2}{RT} \sum_{i=2}^{q} M_i(1 - \bar{v}\rho_1)c_i \qquad \text{(A-28)}$$

With

$$c = \sum_{i=2}^{q} c_i \qquad \text{and} \qquad M_w = \sum_{i=2}^{q} c_i M_i / c$$

$$\frac{1}{rc} \frac{dc}{dr} = \frac{\omega^2 M_w}{RT} (1 - \bar{v}\rho_1) \qquad \text{(A-29)}$$

In another form, analogous to Equation (A-17), it can be shown that $\Delta c/c_0 = \lambda M_w$. Ordinarily the activity coefficients are not unity. Now, association of solutes, solvation, and other interactions which occur in the system produce an apparent result, with the difference between apparent and true weight average results being accounted for in terms of the activity coefficients.

It is sometimes possible to conduct the experiments at the Flory temperature to produce a pseudo-ideal condition in the system, but ordinarily the task of evaluating the nonideality term of Equation (A-25) cannot be avoided.

Sedimentation Transport

According to one definition, the sedimentation coefficient, s, is

$$s_2 = \frac{(dr/dt)}{\omega^2 r} \qquad \text{(A-30)}$$

The term dr/dt really represents the time rate of displacement of a moving boundary, and $\omega^2 r$ is the measure of centrifugal field strength. Thus s is a typical mobility, centimeters per second per unit field, and it has the dimension, second.

Two-Component Systems. Neutral Molecules

Kinetic Analysis

For the earliest description and interpretation of s by Svedberg, a centrifugal force $M(1 - \bar{v}\rho)\omega^2 r$ was equated to a friction force $f(dr/dt)$ in the steady state, with f being a molar friction coefficient. Such a procedure is valid for this two-component, incompressible system in which there is no volume change on mixing; it is a kinetic theory analysis.

Except for the replacement of the statement for driving force by $-d\bar{\mu}/dr$, a procedure which is now in common use, the meaning of s can be suggested. The equation $-d\bar{\mu}/dr = f(dr/dt)$ represents a special case of the generalized expression from the thermodynamics of irreversible processes which is applicable in this simplest of experimental situations. Now, making use of Equations (A-7) and (A-10), and restricting the argument to the region of constant concentration ahead of the moving boundary, we have

$$-f\frac{dr}{dt} = +\frac{d\bar{\mu}}{dr} = \left(\frac{\partial\mu}{\partial r}\right)_c - M\omega^2 r = \left(\frac{\partial\mu}{\partial P}\right)_c \frac{dP}{dr} - M\omega^2 r \qquad \text{(A-31)}$$

Thus, with the application of Equations (A-9) and (A-12),

$$\frac{(dr/dt)}{\omega^2 r} = \frac{M(1 - \bar{v}\rho)}{f} = s \qquad \text{(A-32)}$$

The early combination of s and D to determine molecular weight required the elimination of the molar friction coefficient between this expression and another one for the diffusion coefficient D, which corresponds to

$$D = \frac{RT}{f}\left(1 + c\frac{\partial \ln y}{\partial c}\right) \qquad \text{(A-33)}$$

The result of the combination of the expressions for s and D is well known; it is the famous Svedberg equation. In the limit of infinite dilution,

$$M = RTs_0/D_0(1 - \bar{v}\rho) \qquad \text{(A-34)}$$

This equation is strictly correct when applied to two-component systems in which the measurements of s and D have been carried out at the same temperature, in the same buffer, and when the coefficients s and D have been extrapolated to infinite dilution to give s_0 and D_0. To a certain extent, it is possible to convert values to standard conditions when the experiments must be performed under somewhat different conditions: buffer, temperature, etc.

For the actual experimental evaluation of s, diagrams of either concentration gradient or of concentration versus radial distance in the cell at several different times are required. These diagrams, called boundary gradient or boundary curves, respectively, are obtained directly by the use of certain optical systems, with photography for their registration. Sought from them is the position in the boundary region which moves with the same velocity per unit field as do the molecules which are sedimenting ahead of the boundary. It is the square root of the second moment of the gradient curve, r_{2m}. The equations which are ordinarily used in the evaluation of s depend upon the continuity equation in combination with a flow equation.

Since $J = c\ dr/dt$, the flow in the region of constant concentration ahead of the boundary is $J = cs\omega^2 r$. But in the boundary region itself the flow process involves as well an effect of diffusion. Now,

$$-\frac{d\bar{\mu}}{dr} = -\frac{d\mu}{dr} + M(1 - \bar{v}\rho)\omega^2 r \qquad (A\text{-}35)$$

and, since $\mu = \mu\,(r, P)$

$$J = -D\ dc/dr + cs\omega^2 r \qquad (A\text{-}36)$$

This is the Lamm flow equation, written in conventional form.

To calculate the sedimentation coefficient, operations with the following definitions suffice. From the definition of s, Equation (A-30),

$$dr_{2m}/r_{2m} = s\omega^2\ dt \qquad (A\text{-}37)$$

where r_{2m} is again calculated from the second moment of the boundary gradient curve. When s is constant, $\ln\ (r_{2m}/r_0) = s\omega^2 t$, where r_0 is the distance from the center of rotation to the meniscus. So, from the plot of $\ln\ (r_{2m}/r_0)$ versus $\omega^2 t$, the coefficient s is obtained as the (limiting) slope. Ordinarily, s varies with c (and c varies with time t by radial dilution) so that the simple plot as described is not strictly linear.

Because of the requirement of a sector-shaped cell, there is a radial dilution. For the region ahead of the boundary at some time t, Equation (A-5), with $J = cs\omega^2 r$, becomes

$$\frac{dc_t}{dt} = -2c_t s\omega^2$$

or, for $s = $ constant

$$c_t = c_0 \exp\ (-2s\omega^2 t) = c_0\ (r_0/r_{2m})^2 \qquad (A\text{-}38)$$

It is this radial dilution effect which helps to transform the rate process from one which is simple in physical principle and expression to another which can become enormously complicated in mathematical description.

Nonequilibrium Thermodynamical Analysis

The mathematical problems which appeared when attempts were made to extend the sedimentation velocity equations to polydisperse nonideal solutions turned out to be quite difficult of solution. These extensions really began with Hooyman *et al.* (*1*) when they derived the Svedberg equation within the framework of the thermodynamics of irreversible processes, thus to permit more rigorous definitions of some of the quantities involved. The generalizations for application to the more complicated systems followed. The Lamm "macro-dynamical" treatment, originally considered to be an alternate general approach, has since been shown to be equivalent to the nonequilibrium thermodynamic description.

The transport of molecules in the ultracentrifuge is a rate process, one in which the system is not appreciably removed from equilibrium. Thus, in the thermodynamics of irreversible processes, the flux of the ith component, J_i, is assumed to vary linearly with the forces which cause the flows. Thus, for the flow equations we write

$$(J_i)_a = \sum_k (L_{ik})_a X_k \qquad \text{(A-39)}$$

where the $(L_{ik})_a$ are the phenomenological coefficients for the "a" frame of reference; they are dependent upon the units used to express the forces and the flows. It is required that conjugate forces and flows are used in order to obtain the Onsager relation. Thus they must satisfy the dissipation function

$$T\sigma = \sum_i (J_i)_a (X_i) \qquad \text{(A-40)}$$

where σ is the rate of production of entropy by the irreversible processes.

The subscript "a" is used to emphasize that the choice of reference frame is important. The decision as to whether the Onsager relations will apply for any specified frame depends upon other considerations, as well (*1,2,8*).

If the force is expressed on a per mole basis, the flows must be in moles per square centimeter per second. If the force is set down in this way,

$$X_k = -\frac{\partial \bar{\mu}_k}{\partial r}$$

$$\bar{\mu}_k = \mu_k - M_k \omega^2 r^2 / 2$$

$$\mu_k = \text{chemical potential per mole}$$

$$= \mu_k(C, P) \quad \text{at constant temperature}$$

$$C = \text{concentration on a moles per 1000 cc basis}$$

Then

$$(J_i)_a = \sum_k (L_{ik})_a \left[-\frac{\partial}{\partial r}\left(\mu_k - \frac{M_k \omega^2 r^2}{2} \right) \right]$$

$$= \sum_k M_k (L_{ik})_a (1 - \bar{v}_k \rho) \omega^2 r - \sum_j \left[\sum_k (L_{ik})_a \frac{\partial \mu_k}{\partial C_j} \right] \frac{dC_j}{dr} \quad \text{(A-41)}$$

Actually, such equations are usually expressed on a "per gram" basis and we set down for use the modified forms. The flow on the "per gram" basis, $(J_i)_a'$, is related to the flow on the "per mole" basis, $(J_i)_a$, by the statement $(J_i)_a' = M_i(J_i)_a$. (Quantities on the "per gram" basis are primed.) The total potentials are related, $\bar{\mu}_i' = \bar{\mu}_i/M_i$. Then,

$$(J_i)_a' = \sum_k M_i M_k (L_{ik})_a X_k' \quad \text{(A-42)}$$

$$(L_{ik})_a' = M_i M_k (L_{ik})_a \quad \text{(A-43)}$$

$$(J_i)_a' = \sum_k (L_{ik})_a'(1 - \bar{v}_k \rho)\omega^2 r - \sum_j \left[\sum_k (L_{ik})_a'\left(\frac{\partial \mu_k'}{\partial c_j} \right) \right] \frac{dc_j}{dr} \quad \text{(A-42a)}$$

Thus, on the mole scale,

$$s_i = \frac{1}{C_i} \sum_k M_k (L_{ik})_a (1 - \bar{v}_k \rho) \quad \text{(A-44)}$$

and on the gram scale,

$$s_i = \frac{1}{c_i} \sum_k (L_{ik})_a'(1 - \bar{v}_k \rho) \quad \text{(A-45)}$$

The early kinetic theory equations for s and D, the Svedberg molecular weight formula and the Lamm flow equation, are strictly applicable only to binary systems in which the partial specific volumes of the components are independent of pressure and composition. They did not specify the frame of reference which was used. The newer equations from the thermodynamics of irreversible processes show that the traditional coefficients are those referred to the volume-fixed frame, which with the assumptions which were made, became equivalent to those measured from a mark on the cell. The latter frame is of course the one used by the experimentalist.

With the application of the thermodynamic analysis to the two-component, incompressible system there is found to exist a single flow, described by the Lamm equation, and it is caused by a driving force which is resolved into two components. Therefore, the earlier recourse to the elimination of friction coefficients, assumed to be equal, between the equations for s and D in order to obtain the molecular weight expression is avoided. For the two-component system, with the equations written on the "per gram" scale, the generalized equations written above reduce to the following:

$$s_2 = \frac{L_{22}'(1 - \bar{v}_2\rho)}{c_2} \tag{A-46}$$

$$D = L_{22}' \frac{\partial \mu_2'}{\partial c_2} \tag{A-47}$$

The connection between L_{22}' and the friction coefficient derives from the molecular kinetic interpretation of the sedimentation coefficient,

$$s_2 = \frac{M_2(1 - \bar{v}\rho)}{f_2}$$

It is,

$$\frac{L_{22}'}{c_2} = \frac{M_2}{f_2}$$

The expression for the molecular weight, M_2, is obtained by the elimination of L_{22}' between Equations (A-46) and (A-47) to give, in the infinitely dilute solution,

$$M_2 = \frac{RTs_0}{D_0(1 - \bar{v}\rho)} \tag{A-34}$$

which is again the familiar Svedberg equation.

TERNARY SYSTEMS

The treatment for a solution which contains three components follows the same general outline, but the equations are more complicated. In addition to thermodynamic interactions, one now must contend with couplings of flow. Adequate descriptions for these cases exist, but in recent years a considerably greater effort has been devoted to finding the conditions under which the equations for use with the three-component systems become formally identical with the equations for two-component systems.

Studies of the sedimentation and diffusion of proteins are usually carried out in three-component systems, water (1), protein PX_z (2), and supporting electrolyte, BX (3). With the components defined as indicated, with components 2 and 3 being electrolytes, we have

$$J_2' = c_2 s_2 \omega^2 r - D_{22} \frac{dc_2}{dr} - D_{23} \frac{dc_3}{dr} \tag{A-48}$$

$$J_3' = c_3 s_3 \omega^2 r - D_{32} \frac{dc_2}{dr} - D_{33} \frac{dc_3}{dr} \tag{A-49}$$

The corresponding sedimentation coefficients are

$$s_2 = \frac{L_{22}'(1 - \bar{v}_2\rho) + L_{23}'(1 - \bar{v}_3\rho)}{c_2} \qquad \text{(A-50)}$$

$$s_3 = \frac{L_{32}'(1 - \bar{v}_2\rho) + L_{33}'(1 - \bar{v}_3\rho)}{c_3} \qquad \text{(A-51)}$$

Since the expression for s_2 contains only the two coefficients L_{22}' and L_{23}', we set down the two (of four) expressions for the diffusion coefficients which are involved, one of them the main coefficient, D_{22}, for the protein component and the other its cross-term coefficient, D_{23}, with the electrolyte component.

$$D_{22} = L_{22}'\frac{\partial\mu_2'}{\partial c_2} + L_{23}'\frac{\partial\mu_3'}{\partial c_2} \qquad \text{(A-52)}$$

$$D_{23} = L_{22}'\frac{\partial\mu_2'}{\partial c_3} + L_{23}'\frac{\partial\mu_3'}{\partial c_3} \qquad \text{(A-53)}$$

These two equations descriptive of diffusion behavior are solved to give L_{22}' and L_{23}' in terms of D_{22} and D_{23} and the four $(\partial\mu_i'/\partial c_j)_{T,P,c_k \neq 1}$ for substitution in the expression for s_2, Equation (A-50). The cross-term diffusion coefficients are measurable quantities, as is the sedimentation coefficient, s_2. As expected, s_2 depends also on the concentration and the partial specific volume of the protein component, and it is measurable in a system in which c_2 is kept low. The procedure as described is entirely analogous to that used in connection with the two-component system. For the limiting case (and for which the interaction terms become negligible), as c_2 approaches zero, the result is

$$\lim_{c_2 \to 0} s_2 = \frac{M_2(1 - \bar{v}_2\rho)D_{22}}{RT}\left\{1 + \frac{c_3}{2}\left[\frac{M_3(1 - \bar{v}_3\rho)}{M_2(1 - \bar{v}_2\rho)}\right]\left(-\frac{z}{c_3}\right)\right\}$$

or

$$\lim_{c_2 \to 0} M_2{}^a = M_2\left[1 - \tfrac{1}{2}z\frac{M_3(1 - \bar{v}_3\rho)}{M_2(1 - \bar{v}_2\rho)}\right] \qquad \text{(A-54)}$$

This is exactly the result which was found by Lamm (3) using classical thermodynamics, for the molecular weight to be obtained in the sedimentation *equilibrium* experiment with the system as described. The teaching is again the same: a secondary charge effect remains at zero protein concentration and the molecular weight obtained in the extrapolation will not be exactly the true molecular weight of the protein but somewhat less than this value. However, in most ordinary cases, the correction for the charge term will not exceed a few percent.

Eisenberg (5) has approached the transport problem in the ternary system in another way. The components are defined by reference to a dialysis equilibrium, with the objective of achieving the same kind of simplification in transport as is available at equilibrium. He writes for the flow relative to the cell of the macromolecular component 2 at time t,

$$J_2 = -L_{22}'(\partial \bar{\mu}_2'/dr) \tag{A-55}$$

To obtain the less involved formula, it was assumed that (a) the solution ahead of the boundary is of constant concentration with respect to the re-defined component 2, i.e., $dc_2 = 0$, and (b) the electrolyte component is essentially at equilibrium over the cell, i.e., $d\mu_3' = 0$. With these assumptions the expression for the molecular weight becomes

$$M_2 = \frac{RTs_0}{D_0}\left(\frac{\partial \rho}{\partial c_2}\right)_{P,\mu_3'} \tag{A-34a}$$

This equation is the equivalent of the Svedberg equation, provided the quantity $(\partial \rho/\partial c_2)_{P,\mu_3'}$ is measured in dialyzed solutions. Under the conditions as stated $(\partial \rho/\partial c_2)_{P,\mu_3'}$ becomes very nearly equal to $(1 - \bar{v}_2 \rho_1)$, where ρ_1 is the density of the equilibrium solvent. The coefficient $(D)_0$ could be obtained from observations in the absence of centrifugal field, again at $d\mu_3' = 0$, with $(s_2)_0$ being available in the traditional ways.

The assumption $d\mu_3' = 0$ requires the assignment of the value zero to the term in L_{23}' which would normally have appeared in a second term on the right of the expression for J_2, Equation (A-55). We retain the impression that this simplification may require some further consideration. The cross-term diffusion coefficients may be quite small in comparison with the main ones in ternary systems which contain only neutral molecules, but as of this date there is no assurance that this is true when electrolytes are present.

But by these last statements nothing derogatory is meant or implied. On the contrary, it is a laudable enterprise and from it one begins to see how applicable for the properly conducted experiment the early formulations of Svedberg and of Lamm still are. This circumstance is obviously of great value to the physical biochemist.

References

1. G. J. Hooyman, H. Holtan, Jr., P. Mazur and S. R. de Groot, *Physica* **19**, 1095 (1953); G. J. Hooyman, *ibid.* **22**, 751 and 761 (1956).
2. H. Fujita, "The Mathematical Theory of Sedimentation Analysis." Academic Press, New York, 1962; cf. also J. W. Williams, K. E. Van Holde, R. L. Baldwin and H. Fujita, *Chem. Rev.* **58**, 715 (1958).
3. O. Lamm, *Ark. Kemi, Mineral. Geol.* **17A**, No. 25 (1944).

4. G. Scatchard, *J. Amer. Chem. Soc.* **68**, 2315 (1946).
5. H. Eisenberg, *J. Chem. Phys.* **36**, 1837 (1962).
6. E. F. Casassa and H. Eisenberg, *Advan. Protein Chem.* **19**, 287 (1964).
7. M. Wales, F. T. Adler and K. E. Van Holde, *J. Phys. Colloid Chem.* **55**, 145 (1951).
8. S. R. de Groot and P. Mazur, "Non-Equilibrium Thermodynamics" North Holland Publ., Amsterdam, 1962.

Bibliography

1. T. Svedberg and K. O. Pedersen, "The Ultracentrifuge." Oxford Univ. Press, London and New York, 1940. The classic treatment. It contains much valuable information including now and then an item which is being "discovered" by more recent investigators.
2. O. Lamm, *Ark. Kemi, Mineral. Geol.* **17A**, No. 25 (1944). An early extension, with an important disclosure for the protein physical chemist, of sedimentation equilibrium theory to three-component, strong electrolyte solutions by using a thermodynamic analysis.
3. J. P. Johnston and A. G. Ogston, *Trans. Faraday Soc.* **42**, 789 (1946). Of much interest both for its own particular result and as a pioneer paper on "moving boundary" theory, now applied to electrophoresis and chromatography as well.
4. R. J. Goldberg, *J. Phys. Chem.* **57**, 194 (1953). The first completely general mathematical analysis for multicomponent systems at equilibrium and in transport, the components of which display thermodynamic nonidealities.
5. G. H. Hooyman, H. Holtan, Jr., P. Mazur and S. R. DeGroot, *Physica* **19**, 1095 (1953). Cf. also *Physica* **22**, 751, 761 (1956). Derivation of the Svedberg equation within the framework of the thermodynamics of irreversible processes, pointing the way to generalizations of sedimentation transport theory to multicomponent systems of all descriptions.
6. H. Fujita, *J. Chem. Phys.* **24**, 1084 (1956). Detailed mathematical study of the effects of the variation of apparent sedimentation coefficient as a function of macromolecular solute concentration, with a solution of the Faxén type for the Lamm differential equation of the ultracentrifuge in which allowance for the s/c dependence is made.
7. M. Meselson, F. W. Stahl and J. R. Vinograd, *Proc. Nat. Acad. Sci. U.S.* **43**, 581 (1957). Demonstration that, in the density gradient formed at sedimentation equilibrium in an aqueous heavy salt solution, for a homogeneous macromolecular species, the distribution about its buoyant position is Gaussian, with a standard deviation which is directly related to its molecular weight. It is a very effective experimental arrangement in DNA physical chemistry.
8. J. W. Williams, K. E. Van Holde, R. L. Baldwin and H. Fujita, *Chem. Rev.* **58**, 715 (1958). A theoretical review article, later on broadened and extended by Fujita in his comprehensive monograph, "Mathematical Theory of Sedimentation Analysis," Reference 13 below. The parts descriptive of sedimentation equilibrium in both article and monograph depend quite appreciably on contributions by Wales and associates made during the period 1946–1951.
9. H. K. Schachman, "Ultracentrifugation in Biochemistry." Academic Press, New York, 1959. This is a general and practical discourse on the uses and techniques of ultracentrifugation. It deals with the subject as it was applied to biological systems at the time of publication; it is now somewhat out-of-date, yet it remains a very beneficial source.

10. R. L. Baldwin and K. E. Van Holde, *Advan. Polym. Sci.* **1**, 451 (1960). A sophisticated mathematical review of sedimentation analysis in application to the behaviors in the ultracentrifuge of organic high polymer solutions.

11. S. Claesson and I. Moring-Claesson, *Analy. Methods Protein Chem.* **3**, 121 (1961). Another excellent review, with broad impact and interest because of its considerable attention to experimental detail.

12. H.-G. Elias, "Ultrazentrifugen-Methoden," 2nd ed. Beckman Instruments GmbH, München, 1961. A monograph of high merit in which are set forth the details of the methods of measurement and of the actual numerical evaluation from the data for molecular weights, sedimentation coefficients and related quantities. French (but not English) translation is currently available.

13. H. Fujita, "The Mathematical Theory of Sedimentation Analysis." Academic Press, New York, 1962. The usefulness of this monograph is so great that it can hardly be expressed in this limited space. It provides practical and rigorous solutions to many problems encountered in ultracentrifugation. It shows how precision of experiment may be transformed to accuracy of interpretation.

14. E. F. Casassa and H. Eisenberg, *Advan. Protein Chem.* **19**, 287 (1964). A magnificent theoretical analysis of the thermodynamics of multicomponent systems, this time with particular reference to those solutions of biological interest.

15. J. M. Creeth and R. H. Pain, *Progr. Biophys. Mol. Biol.* **17**, 217 (1967). This review is unique in the way that it deals with different ultracentrifuge methods, their advantages and disadvantages in a given situation. As with the other articles listed here much study and thought have gone into its preparation

APPENDIX B

Molecular Homogeneity and Its Demonstration

The purification and characterization of macromolecules such as proteins are important subjects. The procedures by which molecular heterogeneity may be recognized in the ultracentrifuge and made the subject of quantitative description are understood in principle, but they are often still difficult in practice. Much the same kind of general statements can be made about the establishment of molecular homogeneity for a given preparation—a related problem. The same choice of general approach is available, with both sedimentation transport and sedimentation equilibrium experiments possessing certain advantages. Descriptions of the mathematical analyses involved in tests for molecular homogeneity are presented here, with some mention of experimental results. Again, the treatment is limited in scope, for it deals only with the newer topics and attitudes in the subject with which we have had research experiences.

Sedimentation Velocity

Our own investigations of the subject may be said to have begun with a communication by Baldwin and Williams (*1*) in 1950. They were continued for a decade, largely by Baldwin and by Fujita. Two distinct situations were involved. In the earlier period the emphasis was given to a study of variation of the breadth of the boundary gradient versus distance (time) curve in the sector-shaped cell on transport. The shape of the curve is influenced by diffusion, heterogeneity with respect to sedimentation behavior, and the sedimentation coefficient dependence on solute concentration. The effects of pressure on the sedimentation process, which may be appreciable when organic solvents are involved, were also considered but are not here discussed; we shall restrict the argument to aqueous systems. Guiding principles were found to correct for the broadening of this curve due to diffusion and for its sharpening due to s upon c dependence to give a true distribution of sedimentation coefficients. (Such procedures are illustrated by the text of Chapter II as it pertains to the dextran system.) In the later phases of the program, the same general principles were employed in a test for homogeneity of solute. In transport this test usually depends upon the evaluation of the diffusion coefficient for a *two-component system* as the solute is subject to the driving force, making use of the height/area ratio (H/A) of the gradient curve. An adequate theoretical equation for H/A is therefore required; it has its source in solutions of the Lamm differential equation of the ultracentrifuge.

Most often used of these solutions is that of Faxén. It is an approximate equation, one which was derived by making use of some conditions which are not altogether realistic. However, the solution is simple, and in the early days of ultracentrifugation it provided a very useful method for the analysis of sedimentation velocity data for aqueous solutions of carefully prepared proteins. However, the basic assumptions of Faxén really require that the time rate of displacement of the maximum height of the boundary gradient curve is determined solely by s, irrespective of the value of the diffusion coefficient (2). In practice Svedberg and his associates soon found that by using the analytical expression for H/A from the Faxén solution the values obtained for the diffusion coefficient were uncertain and time-dependent, for reasons not then understood. Thus, it became necessary to measure D in a separate, independent experiment for combination with the s values to obtain molecular weights.

This restriction is now largely removed by the Fujita approximation, one still of the Faxén type. The earlier solution had assumed that both s and D were constants, independent of solute concentration; the newer solution of the Lamm equation allows s to be a linear function of the concentration, but with D remaining constant. It is an observed experimental fact that often s does indeed decrease linearly with concentration or almost so, in the protein systems ordinarily studied. A significant result of the s upon c decrease is a sharpening of the boundary gradient curve and Fujita was able to calculate the magnitude of this sharpening effect. The complete solution is now more complicated, but fortunately a relatively simple relation for the H/A ratio can be taken from it. Thus the evaluation of s and D from sedimentation transport experiments alone now really becomes possible. It turns out that the new procedure gives a value of D which is independent of time, whereas the Faxén result value fell off with time. Provided the solute is homogeneous, a quite correct, time-invariant value of D will be found, thus giving a self-contained procedure for the molecular weight datum. Further, if the apparent D is both time-independent and substantially equal to the coefficient as obtained in the more conventional ways, an excellent test for the homogeneity of a given protein is thereby made available.

THE FAXÉN SOLUTION (3)

In addition to the assumptions that s and D are invariant with solute concentration, the Faxén solution (3) is based upon the two conditions:

$$\varepsilon = 2D/\omega^2 s r_0^2 \ll 1$$

$$\tau = 2s\omega^2 t \ll 1$$

The quantity r_0 is the cell position of the initial sharp boundary formed between solution and solvent. These conditions require that the solution be one for solutes of relatively large size, in other words that the sedimentation transport be relatively uncomplicated by the diffusion process, and that the times involved are small (since $s\omega^2$ should not be too small). The "infinite cell" condition is another restriction. It is

$$c(r, 0) = \begin{cases} 0 & (-\infty < r < r_0) \\ c_0 & (r_0 < r < \infty) \end{cases}$$

Written in the common form, the Lamm equation of the ultracentrifuge is

$$\frac{\partial c}{\partial t} = \frac{\partial}{r \partial r}\left[r\left(D\frac{\partial c}{\partial r} - s\omega^2 rc\right)\right] \tag{B-1}$$

(Again, this equation, written in the above form, is restricted in its application to incompressible, binary solutions. The partial specific volumes of the components are independent of pressure and of composition.)

With the above restrictions its approximate solution may be written as

$$\frac{c}{c_0} = \frac{e^{-\tau}}{2}[1 + \Phi(\xi)] \tag{B-2}$$

where $\Phi(\xi)$ is the error integral and the other newly introduced symbols are dimensionless quantities, as follows:

$$\xi = \frac{1 - [y\exp(-\tau)]^{1/2}}{\{\varepsilon[1 - \exp(-\tau)]\}^{1/2}}$$

$$y = \left(\frac{r}{r_0}\right)^2$$

A number of the higher order terms have been calculated by Gosting (4). Their contributions are of necessity quite small because of the basic assumptions which underlie the solution. Equation (B-2) may be differentiated to give an approximate equation for the concentration gradient distribution, that is, the mathematical description of the form of the so-called boundary gradient curve.

For this curve, the squared height–area ratio is

$$\left(\frac{H}{A}\right)^2 = \frac{\exp(-\tau)}{r_0{}^2 \pi \varepsilon [1 - \exp(-\tau)]} \tag{B-3}$$

In the earlier, more conventional symbols, Equation (B-3) takes the form

$$\left(\frac{A}{H}\right)^2 = D\left\{\frac{2\pi}{s\omega^2}\left[\exp(2s\omega^2 t) - 1\right]\right\} \tag{B-3a}$$

The quantity (A/H) is directly measurable from the boundary gradient curve at the several radial distances, and s is determined as has been indicated in Appendix A. The plot of $(A/H)^2$ versus $(2\pi/s\omega^2)$ $[\exp(2s\omega^2 t) - 1]$ is linear, with slope D, the diffusion "constant." When $2s\omega^2 t \ll 1$, as required by the Faxén conditions

$$(A/H)^2 \cong 4\pi Dt \tag{B-3b}$$

and a knowledge of s is not required. The time t is subject to a zero-time correction.

THE FUJITA SOLUTION (5)

The Lamm differential equation is solved for the case in which s decreases linearly with concentration (but D remains constant)* according to the statement

$$s = s_0(1 - kc)$$

by employing the Faxén approach rather than making the attempt to obtain an exact solution. The results for the c versus r and dc/dr versus r curves are not set down here. We use only the expression for the height–area ratio taken from the latter in the evaluation of D from the sedimentation transport measurements. Following Fujita (5,6) it is

$$\frac{H}{A} = \frac{\exp -\tau/2}{r_0[1 + (1 - \lambda)^{1/2}]\{\varepsilon[1 - \exp(-\tau)]\}^{1/2}} F(\xi_m) \tag{B-4}$$

$$F(\xi_m) = \frac{2\exp(-\xi_m{}^2)}{\pi^{1/2}[1 + \Phi(\xi_m)]} + 2\xi_m$$

$$\xi_m = \frac{\beta\lambda^{1/2}}{1 + (1 - \lambda)^{1/2}}$$

* Recently, Fujita (5a) has relaxed this requirement in further theoretical studies of boundary spreading in transport of polydisperse solutes.

$$\alpha = kc_0$$

$$\beta = (\alpha/\varepsilon)^{1/2}$$

$$\lambda = \alpha[1 - \exp(-\tau)]$$

$$\Phi(x) = \frac{2}{\sqrt{\pi}} \int_0^x \exp(-q^2)\, dq$$

Equation (B-4) can be put in the form

$$\frac{H}{A} \sinh\left(\frac{\tau}{2}\right) = \frac{G(\xi_m)}{2\alpha r_0} \tag{B-5}$$

or, to a good approximation for $\tau \ll 1$, a condition used in the derivation of Equation (B-4),

$$\frac{H}{A}\tau \cong \frac{G(\xi_m)}{\alpha r_0} \tag{B-6}$$

The function $G(x)$ is

$$G(x) = 2x\left\{x + \frac{\exp -x^2}{\pi^{1/2}[1 + \Phi(x)]}\right\}$$

From Equation (B-6) we have

$$\xi_m = G^{-1}[\alpha r_0(H/A)\tau] \tag{B-7}$$

where G^{-1} is the inverse function of G.

Expanding the definition of ξ_m in powers of τ and reverting to the traditional variables we arrive at the statement

$$[1 - \tfrac{1}{2}(1 - kc_0)\omega^2 s_0 t]t^{1/2} = \frac{2D^{1/2}}{r_0\omega^2 s_0 kc_0} G^{-1}\left[2r_0\omega^2 s_0 kc_0\left(\frac{H}{A}\right)t\right] \tag{B-8}$$

which we write as,

$$A = \frac{2D^{1/2}}{r_0\omega^2 s_0 kc_0} G^{-1}(z) \tag{B-8a}$$

In the range of t for which $\tau = 2s_0\omega^2 t \ll 1$, the plot of the quantities A versus $G^{-1}(z)$ should be linear with a slope which involves the diffusion coefficient D. It is seen that its estimation requires advance information of the magnitude of the constants k and s_0; they are quantities made available by a series of sedimentation transport experiments. The height–area ratios are taken as a function of time from the boundary gradient curves. Numerical values of the inverse G function over the proper time interval have been tabulated by Fujita (6).

Making use of the Fujita disclosures of 1956, Baldwin (7) had described a somewhat different approach to the problem, but one still based on the Fujita solution. It made use of a method of successive approximations to determine values of D from H/A measurements, but it also involved the prior knowledge of an approximate D for the system. His computations made use of data taken from sedimentation transport experiments with bovine plasma albumin solutions. In one experiment the value of kc_0 was only 0.066. With the Faxén result the apparent D fell with time to one-half of the known value from experiment, while with the Fujita equations he obtained D values which were not only independent of time but ones which are in good agreement ($\pm 5\%$) with the results of free diffusion experiments of Gosting which are quoted by Baldwin *et al.* (*8*). Fujita's own calculations of the same data also gave very concordant and convincing results (*6*) and one is forced to conclude that the boundary spreading during transport in the ultracentrifuge of an homogeneous solute is very appreciably decreased by even a small s/c dependence.

OTHER METHODS OF ANALYSIS

The study of the Fujita (*6*) and the Baldwin (*7*) reports is a most rewarding experience. Yet there are situations in which somewhat simplified procedures are desirable. Of several such available approaches, two reports are here considered.

In the first of these, Van Holde (*9*) has recast Equation (B-6) into a form which involves as variables only quantities which can be evaluated during a single sedimentation velocity experiment. This is achieved in the following manner.

If the experiment is performed with globular proteins the constant k, and therefore ξ_m, will be small. Then the function $G(\xi_m)$ can be expanded in a Maclaurin series. If the conditions for the experiment are selected so that $\xi_m \leq 0.4$ the first two terms of the series suffice. With $s = s_0(1 - kc_0)$, Van Holde obtained

$$\left(\frac{H}{A}\right) t^{1/2}(1 - \tfrac{1}{2}s\omega^2 t)^{-1}$$

$$= \frac{1}{(4\pi D)^{1/2}} + \left(1 - \frac{2}{\pi}\right)\frac{r_0\omega^2 s}{4D}\left[\frac{kc_0}{1 - kc_0}\right](1 - \tfrac{1}{2}s\omega^2 t)t^{1/2} \quad \text{(B-9)}$$

The plot of the quantity at the left against the variable $(1 - \tfrac{1}{2}s\omega^2 t)t^{1/2}$ for data taken at different times during the performance of the experiment is linear, and by extrapolation to zero time the ordinate intercept gives the quantity $1/(4\pi D)^{1/2}$. The time is measured in seconds.

In Van Holde's report (*9*) is contained a comparison of the D values for bovine serum albumin obtained by Fujita (*6*), and by this shortened analysis; the agreement is extremely good. The Van Holde procedure is very convenient.

The recent report of Karahara (*10*) is descriptive of the use of another simple and approximate equation which has been derived from Fujita's expressions. D is obtained directly from the sedimentation boundary gradient curves. Use is made of the fact that the effect of the s/c dependence on the evaluation of D decreases with decreasing angular speed.

Equation (B-8) is then rearranged to give

$$\left(\frac{A}{H}\right)^2 = 4\pi Dt\left(\frac{4}{\pi}\right)\left[\frac{G^{-1}(z)}{z}\right]^2(1 + s\omega^2 t) \qquad \text{(B-10)}$$

When values for z are in the range $0 < z < 0.16$ the function $(4/\pi)[G^{-1}(z)/z]^2$ can be replaced by $(1 - z)$ and we have the approximation

$$\left(\frac{A}{H}\right)^2 = 4\pi Dt(1 - z)(1 + s\omega^2 t) \qquad (0 < z < 0.16) \qquad \text{(B-10a)}$$

[This statement may be compared to Equation (B-3b). In the limit of vanishingly small k this equation gives $(A/H)^2 = 4\pi Dt(1 + s\omega^2 t) \cong 4\pi Dt$. And, as ω becomes small, whether k is large or not, this simplest of equations should apply.]

Thus, with another suitable choice of experimental conditions the task of evaluating the diffusion coefficient from the form of the moving boundary gradient curve is made appreciably simpler. The experiments were performed in the synthetic boundary cell, to meet more readily the boundary conditions for the mathematical manipulations.

We proceed now with another type of analysis. It has been indicated (*4*) that if the effect of the spreading due to diffusion is too large, actually something one-third to one-quarter of that from heterogeneity, the usual extrapolation procedures will exaggerate the estimate of the solute heterogeneity. Cases in point are found in the Kakiuchi-Williams and Iso-Williams articles, discussed in Chapter IV. In each of these reports, the distributions of sedimentation coefficient data were recognized to be unreliable, and in them expectation of further clarifications and computations was anticipated.

Baldwin (*11*) has examined other methods of extrapolating the boundary gradient curves to infinite time and found that for *Gaussian* distributions of s, extrapolations of the quantity $(S - \bar{s})^2$ at fixed values of $g^*(S)/g^*(S)_{\max}$ versus $\{t \exp(\bar{s}\omega^2 t)\}^{-1}$ had advantages when the effects of diffusion were large. For the Gaussian distribution of s,

$$g(s) = \frac{1}{p(2\pi)^{1/2}} \exp\frac{(s - \bar{s})^2}{2p^2}$$

Here p is the standard deviation and $\bar{s}$ the mean, of the distribution of s.

The function $g^*(S)$ is also Gaussian in form; it is described in the Baldwin (11) article. It derives from boundary spreading in electrophoresis theory. The quantity $g^*(S)$ is obtained directly from experiment by means of Equation (II-18), with S being the reduced coordinate described therein. These curves should be approximately linear, with all the lines passing through the origin.

Results of new sedimentation transport experiments, performed by Hancock (12), for the same multiple myeloma γG-globulin which was used by Kakiuchi and Iso (Chapter IV) were treated by plotting the quantity $(S - \bar{s})^2$ versus $[t \exp (\bar{s}\omega^2 t)]^{-1}$ at several fixed values of $g^*(S)/g^*(S)_{max}$. The solvent was 6 M guanidine hydrochloride, in which the protein neither associates or dissociates. Some of the data points are shown in Figure B-1;

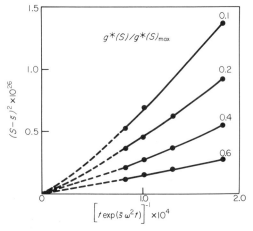

Figure B-1. Plot of $(S - \bar{s})^2$ at fixed values of $g^*(S)/g^*(S)_{max}$ as a function of $[t \exp (\bar{s}\omega^2 t)]^{-1}$ on the *leading* side of the boundary. Solution is of multiple myeloma γG-globulin (HL) at concentration of 0.219 g/dl in 6 M guanidine hydrochloride, and at 25°C. Speed of ultracentrifuge rotor: 52,640 rpm. For $g^*(S)/g^*(S)_{max}$ at 0.2, 0.4, and 0.6 on the *trailing* side of the boundary the plots are substantially alike. (Data of D. K. Hancock.)

they pertain to the leading side of the boundary. Another set of curves, this time for the trailing side of the boundary, not shown, corresponds quite well. Thus, the plots for $g^*(S)/g^*(S)_{max}$ on both sides of the gradient curve at 0.2, 0.4, and 0.6 are in substantial coincidence. The one for this ratio on the trailing side at 0.1 does not pass through the origin, but this is taken to indicate that in the experiment the boundary had not completely broken away from the meniscus. It was considered not to be necessary to make a correction for boundary sharpening due to an s/c dependence because at the low protein concentration the observed s is very close to the zero concentration value, s_0.

With this treatment of the data, the conclusion would appear to be that the myeloma γG-globulin is reasonably homogeneous, as was to have been expected. However, a parallel treatment for the sedimentation transport behavior of this protein when dissolved in pH 7 cacodylate buffer, at $c_0 = 0.855$ g/dl, might seem to lead to the same conclusion, in spite of the fact that under these same solution conditions the data of Kakiuchi require some 18% protein dimerization. On the leading side of the $(S - \bar{s})^2$ versus $[t \exp \bar{s}\omega t)^{-1}$ curve the plots are strictly linear at $g^*(S)/g^*(S)_{max} = 0.1, 0.3, 0.4, 0.5, 0.6, 0.7, 0.8,$ and 0.9, and they all extrapolate well to the origin. On the trailing side all of these curves are concave upward, but they too extrapolate to the origin. Thus, the situation is confused.

Sedimentation Equilibrium

From what has been set down to this point it is apparent that there remain complications in the use of the form and properties of the boundary gradient curve from sedimentation transport experiments in the study of molecular homogeneity, even though this approach may be, in principle, the more advantageous from the point of view of resolution. These complications arise due to the necessity of corrections for diffusion and for concentration and pressure dependencies of sedimentation coefficient.

Turning now to the sedimentation equilibrium experiment, it again appears that further extensions of present knowledge are required. One of the great problems to be solved has to do with the effects of thermodynamic nonideality of the solution in modifying the solute distribution at equilibrium. It is believed to be true that all judgements to date of the sensitivity of sedimentation equilibrium data to solute polydispersity, or lack of it, are applicable only to ideal, two-component solutions, simple solvent and uncharged macromolecular solute. The equilibrium condition is then described by Equation (A-14c), which indicates that a plot of $\ln c$ versus r^2 is linear, and the molecular weight of the solute is available from its slope.

Especially in the recent biochemical literature, linearity of this plot has become a standard criterion of protein homogeneity. Unfortunately, the effect of any thermodynamic nonideality acts to lessen the deviation from the linearity of the curve produced by solute heterogeneity. Thus, this graph may happen to be substantially linear over a considerable range of distance in the cell despite the fact that the solute is quite heterogeneous. There are two other and better tests available (*13*) and a brief description of them is provided for comparison with the less involved and very common procedure. They represent far better probes, but they are still restricted in application to ideal, two-component solutions.

It is readily understood that when the sample is homogeneous, it is also required that:

(a) the quantity $M_{w(r)}$ is independent of r, and
(b) the integral

$$I_{(r)} \left(= \int_{r_m}^{r} rc_{(r)} \, dr \right)$$

is a linear function of concentration, c.

In the Fujita and Williams note (13) the three criteria have the form:

(1) ln c varies linearly with x
(2) $M_{w(x)}$ is independent of x
(3) The integral $I_{(x)}$ is a linear function of c where again

$$x = \text{a reduced coordinate} = \frac{(r^2 - r_a^2)}{(r_b^2 - r_a^2)}$$

$$M_{w(x)} = \frac{(dc/dx)}{\lambda c}$$

$$I_x = \int_0^x c \, dx$$

We reiterate: the experiments must be performed either under ideal or pseudo-ideal solution conditions if these tests are to be applied. The following observations then may be taken as demonstration of the lack of homogeneity in a sample:

(1) curvature of the ln c versus x plot
(2) variation of $M_{w(x)}$ over the cell
(3) deviation of the I_x versus c plot from linearity

Three measures of the deviation were described and analytical expressions were obtained for them. They are defined in the following way:

$$\Delta_1 = \tfrac{1}{2}(\ln c_{x=1} + \ln c_{x=0}) - \ln c_{x=1/2} \tag{B-11}$$

$$\Delta_2 = \frac{(M_{w(x=1)} - M_{w(x=0)})}{\tfrac{1}{2}[M_{w(x=1)} + M_{w(x=0)}]} \tag{B-12}$$

$$\Delta_3 = \left[\frac{(c_{(x=1)} - c_{(x=0)})(dI_x/dc_{(x=0)})}{I_{(x=1)}} \right] - 1 \tag{B-13}$$

Their physical meanings are shown diagrammatically in Figures B-2 to B-4:

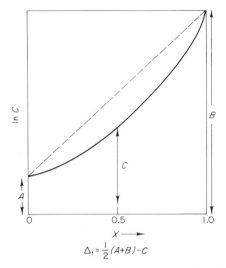

$$\Delta_1 = \frac{1}{2}(A+B) - C$$

Figure B-2. Graphical representation of the measure of deviation, Δ_1. Plot is $\ln c$ versus x. Redrawn from Fujita and Williams (*13*). Reprinted with modification from *J. Phys. Chem.* **70**, 309 (1966). Copyright (1966) by the American Chemical Society. Reprinted by permission of the copyright owner.

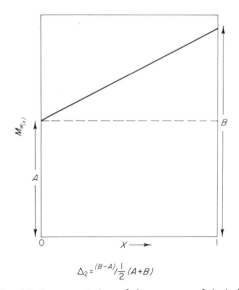

$$\Delta_2 = (B-A) / \frac{1}{2}(A+B)$$

Figure B-3. Graphical representation of the measure of deviation, Δ_2. Plot is $M_{w(x)}$ versus x. Redrawn from Fujita and Williams (*13*). Reprinted with modification from *J. Phys. Chem.* **70**, 309 (1966). Copyright (1966) by the American Chemical Society. Reprinted by permission of the copyright owner.

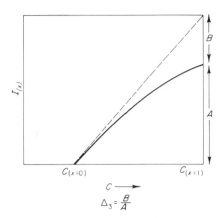

Figure B-4. Graphical representation of the measure of deviation, Δ_3. Plot is $I_{(x)}$ versus c. Redrawn from Fujita and Williams (*13*). Reprinted with modification from *J. Phys. Chem.* **70**, 309 (1966). Copyright (1966) by the American Chemical Society. Reprinted by permission of the copyright owner.

the definitive analytical expressions are available in the original article (*13*). The latter are of a similar general type as the definitions of $P(\lambda)$, $Q(\lambda)$, and $R(\lambda)$ used by Osterhoudt and Williams (*14*) in the presentation of a new method for the estimation of M_w and M_z in polydisperse ideal or pseudo-ideal solutions.

When the experiments are performed under conditions such that λM_w is small, unity or less, and the solute is not too far from being homogeneous, the deviations Δ_1, Δ_2, and Δ_3 are reasonably amenable to numerical evaluation. In a representative case where these experimental conditions were met it was found that Δ_2 and Δ_3 are nearly ten times and five times larger than Δ_1, respectively, thus they represent more favorable and more sensitive tests for homogeneity.

To conclude this section we consider further the meaning of the quantity $I_{(x)}$. It is selected because the quantity $M_{w(r)}$ (or $M_{w(c)}$) has received much the more attention in these seminars. In doing so, we change back to the use of the more classical symbols. By definition, in a polydisperse system, the number average molecular weight at a position r in the cell is,

$$M_{n(r)} = \frac{\sum c_{i(r)}}{\sum c_{i(r)}/M_i} \qquad (B-14)$$

where c_1 is the concentration of the molecules having a molecular weight of M_i. This quantity is proportional to the osmotic pressure at r. The osmotic pressure at this position is equal to the osmotic pressure at the meniscus, r_m, plus the increment in osmotic pressure due to the effect of the centrifugal force on the solute between r_m and a plane at the position r in the cell. Thus,

$$RT \sum c_{i(r)}/M_i = RT \sum c_{i(r_m)}/M_i + (1 - \bar{v}\rho)\omega^2 \int_{r_m}^{r} rc_{(r)}\, dr \quad \text{(B-15)}$$

The first term at the right is a constant of integration, a quantity which can be estimated only on the basis of additional assumptions. The equation can be put in the form:

$$I_{(r)} = \frac{1}{AM_{n(r)}} c_{(r)} - \frac{K}{A} \quad \text{(B-16)}$$

If now, $M_2 = M_n = $ constant, the plot of I_r versus $c_{(r)}$ will be linear.

To illustrate the use of this criterion some unpublished data of Linklater (15) from "low-speed" sedimentation equilibrium experiments are pertinent. The test substances were two Szwarc "living polymer" polystyrenes, the solvent was cyclohexane, and the temperature of the experiment was 34.2°C, the Flory temperature. The $I_{(r)}$ versus $c_{(r)}$ data are given in Figure B-5 for the

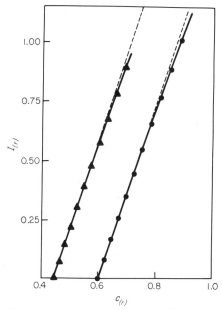

Figure B-5. Plot of $I_{(r)}$ versus $c_{(r)}$ for the "monodisperse" polystyrene for which $M_w = 1.17 \times 10^5$. Solvent is cyclohexane at 34.2°C, the Flory temperature. Initial concentrations: $c_0 = 0.552$ g/100 ml at left, $c_0 = 0.728$ g/100 ml at right. The curve is drawn as a full line through the experimental points. The initial tangent is indicated in the upper portion of each plot. (Data of A. M. Linklater.)

"monodisperse" polystyrene of $M_w = 1.17 \times 10^5$ at two initial concentrations. It is evident that the two curves are slightly concave downward; thus there is a small degree of heterogeneity. For the other polystyrene, of $M_w = 6.00 \times 10^5$, the curves, two initial concentrations, have exactly the same very slight downward concavity, but they are not shown.

Some characteristic constants for these polystyrenes are collected in the following table.

M_w	c_0 (g/100 ml)	M_w/M_n	Slope of full line	Calc. for $M_n = M_w$
1.17×10^5	0.55	1.05	3.675	3.618
1.17×10^5	0.73	1.05	3.588	3.562
6.00×10^5	0.39	1.09	1.225	1.200
6.00×10^5	1.16	1.09	1.227	1.215

For samples of polystyrene which had been fractionated in the usual ways, the downward curvature of the $I_{(r)}$ versus $c_{(r)}$ plots was pronounced.

As an overall statement, it is clear that for the most commonly used plot of $\ln c$ versus r^2, a significant deviation from a straight line will not be observed unless the solute is quite heterogeneous. And, as the preparation of the obvious plot, that of $M_{w(r)}$ versus r, is not at all involved, it is recommended here. Provided the experiment has been performed with $\lambda M_w \cong 1$, the calculation of the quantity Δ_2 is indeed simple and effective in a judgment of solute homogeneity.

References

1. R. L. Baldwin and J. W. Williams, *J. Amer. Chem. Soc.* **72**, 4325 (1950).
2. W. J. Archibald, *Ann. N.Y. Acad. Sci.* **43**, 211 (1942).
3. H. Faxén, *Ark. Mat., Astron. Fys.* **21B**, No. 3 (1929).
4. L. J. Gosting, *J. Amer. Chem. Soc.* **74**, 1548 (1952).
5. H. Fujita, *J. Chem. Phys.* **24**, 1084 (1956).
5a. H. Fujita, *Biopolymers* **7**, 59 (1969).
6. H. Fujita, *J. Phys. Chem.* **63**, 1092 (1959).
7. R. L. Baldwin, *Biochem. J.* **65**, 503 (1957).
8. R. L. Baldwin, L. J. Gosting, J. W. Williams and R. A. Alberty, *Discuss. Faraday Soc.* **20**, 13 (1955).
9. K. E. Van Holde, *J. Phys. Chem.* **64**, 1582 (1960).
10. K. Karahara, *Biochemistry* **8**, 2551 (1969).
11. R. L. Baldwin, *J. Phys. Chem.* **63**, 1570 (1959).
12. D. K. Hancock, unpublished data (1966).
13. H. Fujita and J. W. Williams, *J. Phys. Chem.* **70**, 309 (1966).
14. H. W. Osterhoudt and J. W. Williams, *J. Phys. Chem.* **69**, 1050 (1965).
15. A. M. Linklater and J. W. Williams, Pap., *140th Meet., Amer. Chem. Soc., Chicago, Illinois, September 3–8, 1961* p. 62T (1961).

Author Index

Numbers in parentheses are reference numbers and indicate that an author's work is referred to, although his name is not cited in the text. Numbers in italics show the page on which the complete reference is listed.

A

Adams, E. T., Jr., 19, *20*, 39, 41, 43 (6), 44 (1, 6, 9, 10), 45 (6), *45*, 46 (6), 47 (10, 16), *50*, 53, 55, 58 (9), *61*, *62*
Adler, F. T., 7 (8), 12, *19*, 88, *97*
Alberty, R. A., 26 (12), *35*, 104 (8), *112*
Albright, D. A., 12, *14*, *15*, 19, *20*, 43 (6a), 44(6a), 45, 46, *61*
Archibald, W. J., 100 (2), *112*

B

Baldwin, R. L., 24 (9), 26 (12), *35*, 81 (2), 82 (2), 92 (2), *96*, *97*, *98*, 99, 103, 104, 105, 106, *112*
Bareiss, R., 52 (17), *62*
Bender, M. M., 6 (7), *19*
Bethune, J. L., 39 (3), *61*
Bevington, P. R., 47, *62*
Blair, J. E., 27 (15), *28*, *29*, *35*
Bridgman, W. B., 21, *35*
Buckley, C. E., III, 64 (2), *75*

C

Campbell, D. H., 61 (23), *62*
Cantow, H. J., 27, *35*
Casassa, E. F., 42, 45 (4), *61*, 87, *97*, *98*
Chun, P. W., 44, 47, 60, *61*, *62*
Claesson, S., *98*
Cope, W. T., 44 (10), *61*
Creeth, J. M., *98*

D

de Groot, S. R., 81 (1), 92 (1, 8), *96*, *97*
Deonier, R. C., 10, 11 (12a), 12, *20*, 43 (6, 7, 7a), 44 (6, 7, 7a), 45 (6), *45*, 46 (6), *50*, 51 (7, 7a), *53*, 54, *61*
Donnelly, T. H., 9 (11a), *20*, 22, *35*
Dyson, R. D., 53 (18), *62*

E

Edelman, G. M., 64, *75*
Eisenberg, H., 42, 45 (4), *61*, 87, *97*, *98*
Elias, H.-G., 52 (17), *62*, *98*
Ewart, R. H., 6 (7), *19*, 22 (4), *35*

F

Faxén, H., 101, *112*
Fujita, H., 9 (11), 11 (11, 14, 15), *15*, 19, *20*, 22, 24, 26, *35*, 39, 41, 44, *61*, 65, *75*, 81 (2), 82 (2), 92 (2), *96*, *97*, *98*, 102, 103, 104, 107 (13), 108, *109*, *110*, *112*

G

Gall, W. E., 64 (1), *75*
Gally, J. A., 64, *75*
Gilbert, G. A., 61, *62*
Goldberg, R. J., 7, 8 (9), 11, *19*, 61 (22), *62*, *97*
Gosting, L. J., 24, 26 (12), *35*, 101, 104 (8), 105 (4), *112*
Gralén, N., 26 (13), *35*

Grandine, L. D., Jr., 27, *35*
Green, N. M., 64. *75*
Gross, H., 5, *19*

H

Hancock, D. K., 43 (8), 44 (8), 56, *57*, *61*, 106, *112*
Hess, E. L., 39 (3), *61*
Holtan, H., Jr., 81 (1), 92 (1), *96*, *97*
Hooyman, G. J., 81 (1), 92, *96*, *97*

I

Iso, N., 64 (5), 69, *75*

J

Johnston, J. P., *97*

K

Kakiuchi, K., 64 (4), 67, 68, 70, 71, 72, *75*
Karahara, K., 105, *112*
Kegeles, G., 39 (3), *61*
Kim, S. J., 44 (10), 47 (10), 60, *61*, *62*
Klenin, S. I., *15*,
Kraemer, E. O., 5, *19*, 22, 30, *35*
Kurata, M., *16*, *17*, 20

L

LaBar, F. E., 56, *62*
Lagermalm, G., 26 (13), *35*
Lamm, O., 87, 95, *96*, *97*
Lansing, W. D., 5, *19*, 22, 30, *35*
Linklater, A. M., 19, *20*, 111, *112*
Lys, H., 52 (17), *62*

M

McCormick, H. W., 27, *35*
Mandelkern, L., 12 (16), 18, *20*
Mazur, P., 81 (1), 92 (1, 8), *96*, *97*
Meselson, M., *97*
Moring-Claesson, I., 98

N

Nelson, C. A., 64 (2), *75*
Nichol, L. W., 39, *61*

Nichols, J. B., 3 (2), *19*
Noelken, M. E., 64, *75*

O

Ogston, A. G., *97*
Oncley, J. L., 34, *35*
Osterhoudt, H. W., 9 (10), 19, *19*, 110, *112*

P

Pace, C. N., 46, *61*
Pain, R. H., *98*
Pedersen, K. O., 3 (1), *19*, *97*
Provencher, S. W., 22, *35*

R

Rehfeld, S. J., 34, *35*
Rinde, H., 21, 22, *35*
Rossetti, G. P., 46, 52 (14), 53 (14, 18), 54 (14), *62*

S

Saunders, W. M., 24 (9), 26 (11), *31*, *32*, *33*, *34*, *35*
Scatchard, G., 87, *97*
Schachman, H. K., *97*
Scholte, T. G., 22, *35*
Signer, R., 5, *19*
Singer, S. J., 61 (23), *62*
Squire, P. G., 24 (9), *35*
Stahl, F. W., *97*
Staudinger, H., 4, *19*
Steiner, R. F., 43, *61*
Svedberg, T., 3 (1, 2), 4 *19*, *97*

T

Tagata, N., *16*, *17*, 20
Tanford, C., 64 (2), *75*
Tang, L.-H., 44 (10), 47 (10), *61*
Thompson, J. O., 6 (7), *19*, 22 (4), *35*
Timasheff, S. N., 45, 46 (11, 12) *61*
Tiselius, A., 39, *61*
Townend, R., 45, 46, *61*

U

Utiyama, H., *16*, *17*, 20

V

Valentine, R. C., 64, *75*
Van Holde, K. E., 7 (8), 10, 12, *19, 20*, 46
  52 (14), 53 (14, 18), *54, 62*, 81 (2), 82 (2)
  88, 92 (2), *96, 97, 98*, 104, 112
Vinograd, J. R., *97*
Visser, J., 43 (6), 44 (6), 45 (6), *45*, 46, *50, 61*

W

Wales, M., 6, 7, 12, *19*, 22, 34, *35*, 88, *97*
Weissberg, S. G., 12 (16), 18 (16), *20*

Williams, J. D., 44 (10), 47 (10), *61*
Williams, J. W., 6 (7), 9 (10), 10, 12, *14, 15*,
  19, *19, 20*, 22 (4), 24 (9), 26 (11, 12), 27
  (15), 28, 29, *31, 32*, 33 (19), *33, 34, 35*, 43
  (6, 6a, 7, 8), 44 (6, 6a, 7, 8), 45 (6, 6a),
  45, 46 (6, 6a, 7), 47, *50*, 51 (7), *53*, 54 (7),
  56, *57*, 61 (22), *61, 62*, 64 (4, 5), 67, 68,
  69, 70, 71, 72, *75*, 81 (2), 82 (2), 92 (2),
  96, 97, 99, 104 (8), 107 (13), 108, *109*,
  110 (14), *110*, 111 (15), *112*
Williams, L. C., 12 (16), 18 (16), *20*
Winterbottom, R. J., 46 (12), *61*

Subject Index

Accuracy
  of equilibrium constant data, 45
  of protein concentration data, 46
Analytical gel chromatography, 39
Antibody activity, 64
Antibody fragmentation, 64
Antigen–antibody reaction, 61
Association constant (equilibrium constant), 43, 45, 59
Association mechanisms, 39, 40, 41
  monomer–dimer, 40, 44, 47, 51, 54, 55, 60, 65
  monomer–dimer–trimer, 58
  random (indefinite, isodesmic), 40, 51, 53
Average concentration (c_{av}), 65
  numerical average ($\bar{c}$), 8, 65
Average deviation in $M_{w(c)}^a$, 45

Boundary gradient curve (in transport), 23, 26, 28, 33, 91
  maximum height, 28, 65
  second moment, square root (r_{2m}), 23, 28, 91
Buffer systems
  NaCl–glycine (pH 2.64), 43
  NaCl–cacodylate (pH 7.0), 67ff.
  Urea (8 M), 67ff.
2-Butanone, 16

Cells, shape of, 80
Centrifuge, optical, 3, 5
Centrifugal potential, 82
Charge effect, 87, 95
Chymotrypsinogen A, 40, 43, 55ff.
Colloidal solution, 3
Combination of s and D for M (Svedberg equation), 94

Compressibility, 5, 24
Concentration
  correction for diffusion, 23
  correction for pressure, 24
  correction for solute concentration, 25
  dependence of sedimentation coefficient on, 23, 65ff.
Conditions for sedimentation equilibrium, 82, 83
Conservation of mass, 80
Continuity equation, 25, 81
Conversion
  of s to standard conditions, 90
  of D to standard conditions, 90
Countercurrent distribution, 39
Coupled flows, 94
Cyclohexane, 28, 29

Density, 6, 24, 41
  variation with pressure, 24
  with concentration, 61, 96
Dextran, 22, 30
Diffusion, 99ff.
  correction for, in heterogeneous system, 23
Diffusion coefficient, 90
  apparent value, 21
  corrections for, 23, 90
  determination, 100ff.
Dimensionless distance parameter, 9
Dimerization constant, 51
Dissociation constant, 50
  gram basis, 50
  mole basis, 50
Distribution function, 22
  apparent value, 23
  differential form, 27

Distribution functions (*continued*)
  integral form, 27
  for molecular weight, 26ff.
  for sedimentation coefficient, 23ff., 32ff.
Donnan effect, 87

Electrical potential, 82
Electrophoresis, 39, 61
Enthalpy change, 50
Entropy change, 50
Equilibrium constant, 42
  monomer–dimer (K_2), 40ff.
  monomer–dimer–trimer (K_3), 55
  random, intrinsic constant k, 40ff.
Equilibrium dialysis, 96
Ethyl n-heptanoate, 12
Excess chemical potential, 42
Excluded volume, 18
Extinction coefficient, 46

Faxén solution, 100, 101
Fragment
  F_{ab}, 64
  F_c, 64
Flexible coil molecule, 34
Flory temperature, 6, 12, 18, 22, 26, 89, 111
Flow, 91
  definition of, 91
  equation for, 91
Fraction of monomer in equilibrium mixture, 43, 48
  apparent value of, 47
  true value of, 47
Frame of reference, 92ff.
Free energy change (Gibbs), 49
Fujita solution, 102ff.
γ-G-globulin subunits, 64, 69, 73, 74
γ-G-immunoglobulin structure, 64
Gaussian distribution of s, 105, 106
Geometry of solution cell, 80
Gibbs–Duhem relation, 85
Gibbs equation, 82
Gibbs free energy, 49
Gilbert theory, 61

Height–area ratio, 100ff.
Heterogeneity
  of M, 4, 5, 21, 22, 99
  of s, 23ff., 105
Homogeneity, 41, 79, 99ff.

Ideal solution, 7, 8, 39
Immunoglobulin, 64
Incompressible binary system, 6
Intrinsic association constant, 40, 51, 53, 55ff.
Iso-octane, 12

Johnston–Ogston effect, 26, 97

Kinetic theory approach, 5, 81, 90

β-Lactoglobulin B, 40, 43, 44ff.
Lamm differential equation of the ultracentrifuge, 100, 101, 102
Lamm flow equation, 91, 93
Least squares analysis (cf Evaluation of K_2 and BM_1), 47, 48
Light scattering, 6, 9, 18, 19, 39, 40, 44, 46, 84, 85
Low-speed (short column) method for M, 8, 18, 41, 85
Lysozyme (muramidase), 40, 43, 51ff.

Macro-ion, 6, 82, 86, 87, 95
Macromolecule, 4
Meniscus depletion (high speed) sedimentation equilibrium experiment, 8, 43
Molar friction coefficient, 24, 90, 94
  pressure correction, 24
Molecular models for γ-G-globulin, 64
Molecular weight averages over the cell
  M_n, M_n^a, M_w, M_w^a, M_z, 22, 29, 30, 41, 108, 110
  averages at fixed concentration, $M_{n(c)}$, $M_{n(c)}^a$, etc., 41ff., 65ff.
  distribution of, 26, 27
  most probable value, 31
  most probable distribution, 29
Monomer molecular weight, 42, 43, 44ff.
Multicomponent systems, 85, 87, 92ff.
Myeloma γ-G-globulin, 63ff. 106

Nonequilibrium thermodynamics, 92ff.
Nonideality of solution behavior, 4ff., 39ff.
Nonideality terms, 40ff.
Nonuniformity coefficient, 31
Normalization of data, 30, 32
Normalized weight frequency function, 23

Oligostyrene, 15
Optics
  Interference (cf Rayleigh interference optics)
  schlieren, 27, 66
Organic high polymers, 5, 79
Osmotic pressure, 6, 18, 39, 40, 84
  second virial coefficient, 7, 84
Onsager relations, 92

Pair-thermodynamic interaction parameter, 19
Partial specific volume, 6, 24, 40, 41, 56, 83
  buoyancy term, $(d\rho/dc)_\mu$, 96
  pressure correction, 24
Phenomenological coefficients, 92
Plasma extender, 30
Plateau region, 24, 91
Polydispersity (cf heterogeneity), 6, 21, 79
Polyisobutylene, 12, 18
Polystyrene, 13, 22, 28, 29
Potential
  chemical, 82, 83, 87
  electrical, 82
  total, 82
Pressure
  correction for, 24, 84
  variation with radial distance, 25
Probability integral, 101, 103, 105
Pseudo-ideal solution, 18, 89, 108
Purity of component (cf homogeneity)

Quaternary structure of proteins, 39, 74

Radial dilution law, 25, 91
  modification for pressure effects, 25
Radial distance, 6, 80
  dimensionless parameter, 9
Random-coil macromolecules, 34
Rapidly reversible self-association, 40ff., 63ff.
Rayleigh interference optics, 13, 17, 43, 44, 66
Reciprocal plot, 15, 42
Refractive index increment, 40, 56, 61
Rotor speed, 41
  variations of, 7

Sedimentation coefficient, 24, 25, 26, 65ff., 89, 90
  charge effects, 95
  concentration dependence, 65, 66, 68
  corrections, 25, 32
  determination of, 91
    species in equilibrium mixture, 68, 72
  distribution of, 26, 27, 28
  pressure effect, 25
  variation with radial distance, 25
Sedimentation equilibrium, 3, 4, 46, 63, 65, 81, 107
  charge effect, 86, 87
  ideal behavior, 83
  kinetic theory, 81
  thermodynamic theory, 82
Sedimentation transport, 3, 4, 89ff.
  kinetic theory, binary systems, 90, 91
  thermodynamic theory, multicomponent systems, 92ff.
Self-association of proteins, 39ff.
Skewed peaks, 67
Specific refractive increment, 32, 61
Subunits in proteins, 39, 74
Super-position of $M^a_{w(c)}$ vs c data, 41, 53. 58
Svedberg equation, 90, 92, 94

Temperature dependence of equilibrium constant, 50
Ternary systems, 85ff., 94ff.
Theory of ultracentrifugal analysis (Appendix A), 79ff.
Thermodynamics of irreversible processes, 92ff.

Ultracentrifuge cell, 80

Valence of macro-ion, 82, 86, 87, 95
Van't Hoff plot, 50
Viscosity, 90

Weight average molecular weight
    over the cell, 9ff.
  apparent value, 7, 17ff.
  at fixed concentration, 41ff.
  apparent value of same, 41ff.